Beckett, Royce.
Numerical calculations and algorithms

Numerical Calculations and Algorithms

McGraw-Hill Series in Information Processing and Computers

J. P. Nash, *Consulting Editor*
Richard W. Hamming, *Consulting Editor*
Norman Scott, *Consulting Editor*

Beckett and Hurt, Numerical Calculations and Algorithms
Conte, Elementary Numerical Analysis
Davis, Computability and Unsolvability
Evans and Perry, Programming and Coding for Automatic Digital Computers
Gotlieb and Hume, High-speed Data Processing
Humphrey, Switching Circuits: With Computer Applications
Karplus, Analog Simulation
Ledley, Programming and Utilizing Digital Computers
Levine, Methods for Solving Engineering Problems
McCormick, Digital Computer Primer
Oakford, Introduction to Electronic Data Processing Equipment
Rogers and Connolly, Analog Computation in Engineering Design
Scott, Analog and Digital Computer Technology
Southworth and De Leeuw, Digital Computation and Numerical Methods
Weiss, Programming the IBM 1620: The Hands-on Approach
Williams, Digital Computing Systems
Wrubel, A Primer of Programming for Digital Computers

Numerical Calculations and Algorithms

Royce Beckett, professor of mechanics, engineering college, university of iowa

James Hurt, instructor, engineering college, university of iowa

McGraw-Hill Book Company, new york, st. louis, san francisco, toronto, london, sydney

Numerical Calculations and Algorithms

Library of Congress Catalog Card Number 66-24459

04250

1234567890 MP 7321069876

Preface

Written for a first course in numerical calculations for engineering and science students, this book has evolved from notes used in a course offered to seniors and first-semester graduate students at the University of Iowa. The course requisites are completion of differential and integral calculus. Some experience with ordinary differential equations is helpful but is not required.

The intent is to present numerical methods for solving typical problems in engineering and science on a digital computer. The general objective in each method is to arrange the problem into a sequence of computations and decision steps that can be performed on any modern digital computer system. For this purpose a symbolic language is defined in Chapter 2. The symbolic-language statements are used in a flow chart describing the operations and the order of their performance. Preparation of a flow chart is the objective for each of the numerical procedures presented. The detailed explanation of procedures and the simple example problems used for illustration make the book suitable for self-study.

The use of a computer for solving some of the problems is highly desirable for purpose of illustration and provides high motivation for the student. This is accomplished by the introduction of appropriate program statements to do the operations defined in the flow chart. Manuals ordinarily available at a computer facility will give sufficient guidance for converting the flow charts to program statements. At Iowa the practice has been to prepare some five or six punched-card programs in Fortran for the IBM 7044 system. Prior experience with Fortran, although desirable, is not required.

Since not all the topics in the book can ordinarily be covered in one three-semester-hour course, there is room for choice of those methods which best suit the needs and background of a given class.

The authors are pleased to acknowledge the assistance of former staff member John Denkmann in preparing notes from which the manuscript evolved. The invaluable review of the manuscript by Dr. Richard Hamming, of Bell Laboratories, is gratefully acknowledged. Much credit is given to Mrs. Marilyn McAlexander and Mrs. Lois Adams for their diligence in typing the manuscript and last but not least to our wives Shirley and Gretchen for their forbearance.

Royce Beckett

James Hurt

Contents

Chapter 2 The Flow Chart 15

Chapter 3 Nonlinear Algebraic Equations 43

Chapter 4 Simultaneous Linear Equations, Determinants, and Matrices 81

Numerical Calculations and Algorithms

Chapter 1

Introduction to Computers

1.1 Introduction This book presents some of the more pertinent methods for solving scientific and engineering problems on a digital computer. The methods have been chosen on the basis of their usefulness to the scientist and engineer and of their adaptability for machine computation. The objective for each problem studied is the preparation of a step-by-step procedure, called a *flow chart*, that can be readily translated into a machine program. For a full appreciation of the techniques employed a general understanding of a computer and its capabilities is essential. To this end, the numerical operations of which a modern high-speed digital computer is capable are examined, and some attention is called to the basic components of the computer and their function in problem solving.

1.2 Capabilities of a Computer

This discussion of computer capability includes only those things which bear directly on the preparation of programs, and it is sufficiently general to apply to any high-speed digital computer.

A digital computer can perform the basic arithmetic operations of *addition*, *subtraction*, *multiplication*, and *division*. All arithmetic computations to be performed by the computer must be presented to it in a sequence of coded instructions involving only these basic arithmetic operations. This sequence of instructions is called a *program*. Each computation a computer performs requires an instruction for its execution, and since even a simple problem may involve several thousand calculations, the number of instructions to be executed is very large.

Two characteristics of the digital computer make it possible to perform the same or similar instructions over and over again, so that a very long list of computations can be made with only a few instructions. One of these characteristics is the machine's ability to *identify a number as positive, zero, or negative*, and the other is its ability to *change the operand in an arithmetic computation by altering an instruction internally*. Both these operations are controlled by the program.

The ability of the computer to identify the sign of a number is referred to as the *logic capability*. The number may be a preassigned datum, or it may be the result of some computation. The computer can test the number and identify it as negative (less than zero), zero, or positive (greater than zero). On the basis of this test, alternate paths in the program are chosen. Each test admits three possible courses of action. If the number is negative, a certain specified sequence of instructions will be executed; if it is zero, some other sequence of instructions; and finally, if the number is positive, still another sequence.

The ability of a computer to change the operand in an instruction internally rests on the fact that the location of the operand in the computer is represented by a number code. A number can be added to, or subtracted from, an instruction, thereby changing the instruction so that a different operand will be used when it is repeated. This ability, along with the logic capability, enables the computer to repeat instructions over and over but with different operands. This characteristic is essential in the operation of a computer because most numerical methods are based on iterative routines that perform a series of computations over and over again, using different operands in successive iterations.

Other important considerations in programming that are not directly concerned with computations include getting the program and initial data into the machine, storing the program and data so that they will be available when needed, and getting the results of the computation out of the machine in legible form. Little is said here about these important functions, as they vary greatly from machine to machine. Instead, it is assumed that appropriate elements are available to handle input and output and that adequate space is available for storing the program and all necessary data.

1.3 Components of a Computer

It is helpful when programming to have some knowledge of the functional components of a computer and to know something about the flow of data and instructions in the machine. The basic components of a computer can be summarized according to their function.

1. An *input unit* is capable of reading information into a *storage unit*, which is referred to as the *memory* of the computer. This

information, in the form of data or instructions, can be on punched cards, punched tape, magnetic tape, or any other medium acceptable to the particular input device.

2. A *storage unit,* or *memory,* stores the program, data, and intermediate results where they are available when needed by the computer.

3. An *arithmetic and control unit* directs all the operations of the computer according to the program and performs the computations. This unit includes the logic and instruction-modification capabilities of the computer.

4. An *output unit* displays the results legibly.

Figure 1.1 shows these components schematically. The solid lines indicate the directions of flow for the data and coded instructions, and dashed lines indicate the direction of flow for the control signals.

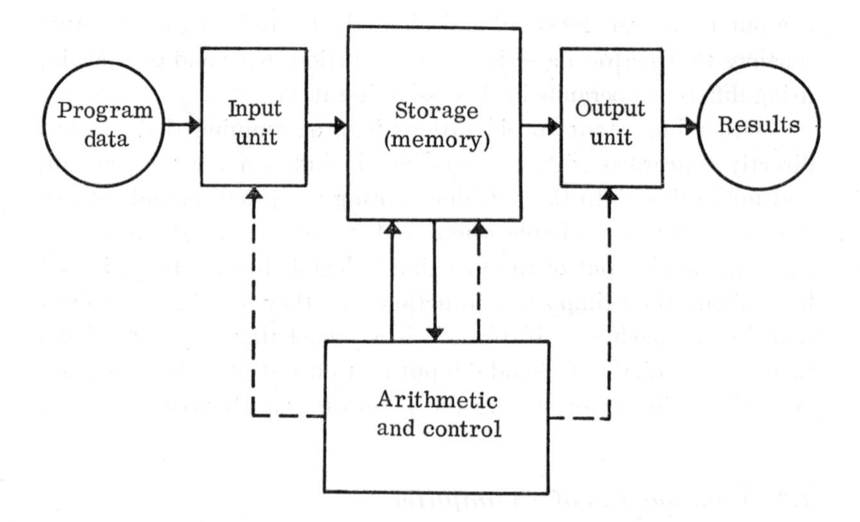

Figure 1.1 Schematic drawing of a digital computer.

Information (instructions or data), properly coded so that the machine can interpret it, is introduced through the input unit of the machine. On command of the control unit, this information is transferred to the memory, where it is stored, and where it is available to the arithmetic and control unit of the computer.

The memory unit is broken up into many cells called *words*. Each word contains a sequence of numerical characters representing a numerical datum or a coded instruction. Each word is given an address, and the information in the word is summoned by that address. The information stored in a word may be called as many times as required in the program, remaining unchanged in that word until it is replaced by the program with another piece of information. The data in a word in memory can be read out as often as necessary without disturbing the contents of the word. They will be destroyed, however, when another piece of information is stored in that word. This concept, called *nondestructive read-out and destructive read-in*, is an important consideration when preparing a program for the machine.

The coded instructions are transferred one by one from memory to the control unit, where they are interpreted and the proper commands sent to the machine. If the instruction indicates an arithmetic operation, the necessary data for the computation are called from the memory, the calculation is made, and the results stored back in the memory. If the instruction indicates that data are to be read into memory, the control unit gives the proper commands to transfer information from the input unit into memory. If the instruction indicates that data are to be displayed as output, the control unit gives the proper commands to transfer information from memory to the output unit, where the information is displayed in some legible form.

To summarize: Information in the form of coded instructions and data is brought into memory via the input unit. From memory the information can be transferred to the arithmetic and control unit for processing. The results must be stored back in memory, from which they and any other information in memory can be transferred as directed by the control unit to the output unit for display.

1.4 Language of Computers

Digital computers store and handle information by coding it in a sequence of positive or negative (yes or no) states of a basic storage element, e.g., an electron tube which is conducting or not conducting, a capacitor which is charged or not charged, or a magnetic element which is polarized north-south or south-north. The identifying state of the element must be capable of being changed easily and quickly and of being sensed, so that stored information can be read by the computer. For reasons which will be evident later, let one state of the memory element be denoted by 0 and the alternate state by 1, e.g., an element magnetized NS is interpreted as 0 and the SN state as 1. If a group of such elements is placed sequentially, they can be made to represent information. For example, if the sequence 01 represents "add," when the computer finds this symbol in the operation portion of a coded instruction, it will add.

A coded instruction contains an instruction number, which is the address or location of that instruction in memory, an operation code, and an address which identifies the location in memory of the operand. Each piece of information is coded in 0's and 1's and appears in memory as a sequence of memory elements in the 0 or 1 state.

Arithmetic numbers, called *data*, are also represented by the symbols 0 and 1. The binary number system, which is a number system referred to the base 2, uses but two characters to represent any integer, viz., 0 and 1. Just as any number can be represented in the 10's digit system by coefficients on successive powers of 10, so any integer can be represented in the binary system by coefficients on successive powers of 2. An example will illustrate. Consider the number 231, which is an abbreviation for

$$231 = 200 + 30 + 1 = 2 \cdot 10^2 + 3 \cdot 10^1 + 1 \cdot 10^0$$

The numbers 2, 3, and 1 are coefficients on successive powers of 10. Whereas 10 digits, 0 to 9, are required to represent numbers to the base 10, only two digits, 0 and 1, are required to represent numbers in the binary system. The number 231 is readily verified to be the sum of the following sequence of powers of 2:

$$1 \cdot 2^7 + 1 \cdot 2^6 + 1 \cdot 2^5 + 0 \cdot 2^4 + 0 \cdot 2^3 + 1 \cdot 2^2 + 1 \cdot 2^1 + 1 \cdot 2^0$$

The coefficient of each successive power of 2 is either 0 or 1 and, as it was for the base 10, the number is represented by the sequence of coefficients. Thus 231 is written in binary form as 11100111.

Table 1.1 gives the binary form for the numbers 0 through 10.

Table 1.1
Binary Numbers

	Binary form	Decimal number
$0 \cdot 2^0$	0	0
$1 \cdot 2^0$	1	1
$1 \cdot 2^1 + 0 \cdot 2^0$	10	2
$1 \cdot 2^1 + 1 \cdot 2^0$	11	3
$1 \cdot 2^2 + 0 \cdot 2^1 + 0 \cdot 2^0$	100	4
$1 \cdot 2^2 + 0 \cdot 2^1 + 1 \cdot 2^0$	101	5
$1 \cdot 2^2 + 1 \cdot 2^1 + 0 \cdot 2^0$	110	6
$1 \cdot 2^2 + 1 \cdot 2^1 + 1 \cdot 2^0$	111	7
	1000	8
	1001	9
	1010	10

Any integer can be represented in the binary number system by a sequence of 0 and 1. If the storage unit of a computer is a sequence of elements that can be put into either of two states, then it can handle, without any confusion, numerical data that are represented in binary form. Coding of computer operations in the binary digits enables the computer to store and properly identify the operation to be performed.

Basic operations in modern digital computers are carried out using the binary code. However, the conversion to binary form is done internally and is automatic. Numerical information, either in the form of a coded program or data, can be presented to the machine in decimal form. The coded program written in decimal form is referred to as the *machine-language program.* An example of machine-language instructions is illustrated in Table 1.2.

The address of the instruction gives its location in memory. The instruction contains the operation code and the address in memory of the operand. Let the operation code 08 be "clear and add to the arithmetic unit," 12 be "add," 91 be "store," and 10 be

Table 1.2
Machine-language Instructions

Address of instruction	*Instruction*
1230	08 2426
1231	12 2123
1232	91 0641
1233	10 3299

"transfer control." This list of instructions will be performed sequentially, and the following operations will be done.

The instruction in the address 1230 tells the computer to clear the arithmetic unit of whatever is there and replace it with the contents of location 2426. Next, instruction 1231 is executed, which says to add to the contents of the arithmetic unit the contents of location 2123. This gives the sum in the arithmetic unit. Instruction 1232 stores the result in location 0641. Instruction 1233 is a transfer-control operation, which says "go to location 3299 for the next instruction." Such basic and detailed instructions are essential for the operation of current digital computers.

Since writing a program in this language is extremely tedious and difficult, much effort has been spent in developing easier methods for writing computer programs. Among the results are such problem-oriented languages as Fortran, Algol, Cobol, Mad, and others, which essentially make the computer write its own machine-language instructions from problem-oriented statements. The computer program that translates the problem-oriented statements into machine-language instructions is called a *compiler*. The compiler accepts general statements and generates the appropriate machine-language instructions. Writing a program in a problem-oriented language is generally a much simpler task than writing a program in machine language.

1.5 Integers and Floating-point Numbers

The way in which integers are represented in the binary number system has already been shown. Fractional numbers can also be written in binary form in a way that is similar to the tens system. The number 1.75 is equivalent to

$$1 \cdot 10^0 + 7 \cdot 10^{-1} + 5 \cdot 10^{-2}$$

In the binary code this number would be represented by

$$1 \cdot 2^0 + 1 \cdot 2^{-1} + 1 \cdot 2^{-2} = 1 + \tfrac{1}{2} + \tfrac{1}{4} = 1.75$$

If one used a decimal point between the 0 power and -1 power of 2, as in the decimal system, the number 1.75 would be written 1.11 in binary form.

Such a system might be used for storage of fractional numbers, called *floating-point* numbers, but it would impose serious limitations on the size of numbers that could be handled. For example, to represent the number 10^{-20} in decimal form requires a decimal point and 20 digits. Many more digits are required in the binary system. However, if the number is written in exponential or scientific form, only a few digits are required. Thus

Number	*Exponent*	
1	-20	$= 10^{-20}$

This is a great economy in storage space.

Floating-point numbers are represented in the computer in exponential form. Clearly arithmetic operations with floating-point numbers are very different from the arithmetic with integers because of the different meaning attached to locations in the memory. An example using decimal numbers will illustrate the distinction between the integer and floating-point modes. The number 20 is represented in integer and floating-point modes as follows:

+00 20	Integer (20)
+0 .. 2 + 20 ... 0	Floating point (20)

The floating-point number is represented by an exponent and the mantissa of the number. The exponent is the integer 2, which appears at the front of the word. The mantissa, which represents the decimal digits in the number, is given in the last portion of the word. A decimal point in front of the mantissa is implied. Thus the floating-point notation represents the decimal +0.20 ... 0 multiplied by 10 raised to the integer power +0 .. 2. A term-by-term addition of the digits in the example above would produce nonsense.

Since scientific calculations generally are carried out in the floating-point mode, that will be assumed to be the normal mode of operation in this text. However, many control statements on digital computers *must* be in the integer mode, and on occasion there is a need for integer arithmetic. Locations in memory are always in the integer mode; control for iteration computations is often in the integer mode; and identification of vector and tensor quantities by subscripting is done in the integer mode. Thus almost every program needs both integer and floating-point numbers. The computer must have some signal for the identification of the mode of a number so that it can be stored in its proper form. In this text, the variables that are integers are listed as such for each problem. If a variable is not listed as an integer, it is assumed to be a floating-point variable.

1.6 Variables

The identification of numerical information in the computer is made by its location in memory. Each location has a number, called an address, which is used to call the information. This is the operational, or machine-language, level of the program. The problem-oriented languages used for coding scientific problems identify data by name, and the compiler allocates a space in memory for the data called by this name. In the course of the execution of the program the content of this space is the current value of the quantity assigned to the location. Quantities identified by name are called *variables*, and the name given to the quantity is the *variable name*.

Variable names may be a single alphabetic character or a combination of alphabetic and numeric characters, but the first character must *always* be alphabetic. In conformity with usual practice, all alphabetic characters are uppercase. Some examples of variable names are listed below. (Note the slash through the letter *O*, which is used to distinguish the letter *O* from zero.)

A	X15	UPSET	VECTØR
SUM	MA7	INDEX	TEMP

Suppose a datum is assigned the name A by the programmer, who is coding in a problem-oriented language. The compiler, which translates the program to machine language, assigns a location in memory to the variable identified by A. When A occurs again in the program, the compiler recognizes it as referring to the contents of the location in memory that is assigned to A and sets up the appropriate machine-language instructions. This represents a great simplification in coding, because data can be handled in arithmetic computations by simple algebraic formulas in which the numerical information is represented by the usual algebraic names. Such statements as

(1) A + B
(2) X * PI/2.5

have the usual everyday meaning and are recognized and translated as such by the compiler. Statements (1) and (2) are translated into machine-language instructions that

1. Add the contents of location named A to the contents of location named B.
2. Multiply the contents of location named X to the contents of location named PI and divide the result by 2.5.

1.7 Subscripted Variables

An important capability of computer language is the use that can be made of subscripting. Vector and matrix elements are normally identified by subscript notation. Thus the vector A is represented by its n components, $a_1, a_2, a_3, \ldots, a_n$. Likewise, the elements of a k by l matrix B are represented by the subscript notation

$$\begin{matrix} b_{11} & \ldots & b_{1j} & \ldots & b_{1l} \\ b_{21} & \ldots & b_{2j} & \ldots & b_{2l} \\ \ldots & \ldots & \ldots & \ldots & \ldots \\ b_{k1} & \ldots & b_{kj} & \ldots & b_{kl} \end{matrix}$$

The elements of the vector A and the matrix B can be stored in the memory of the computer and identified by subscript notation. The vector A would be stored in locations identified by name and subscript. Thus the first element of the vector A might be stored in a location called AVEC(1), the second in AVEC(2), and so on to AVEC(N). The name AVEC is an arbitrary choice. Any name satisfying the conditions in Sec. 1.6 can be used. Each element is called by its name and subscript. For example, the fifth element of A would be called by using AVEC(5) in an arithmetic statement, the ith element by AVEC(I). Similarly the matrix elements are called by name and subscript. The element in the ith row and jth column of the matrix B would be called by the name B(I,J).

By using variable names for the subscripts such as I for the vector (I, J for the matrix), it is possible to cause the program to pick different elements of the vector (or matrix) by simply altering the value for I (I and J).

1.8 Special Functions

Most computer facilities have written program routines that are available to the programmer for evaluating common functions that recur many times. If a routine for a special function is available, it can be called by name whenever it is used. When one of these functions is encountered during the execution of a program, control is automatically transferred to a special routine for evaluating the function.

In this text it is assumed that certain function routines, generally available in a good computing facility, can be called by name. The functions and the notation used are given in Table 1.3. The argument x of these functions may be a number, a variable name, or an arithmetic expression that yields a number. If a function that is not available for call by the program is used, it must be computed separately.

Raising a number to a power is considered an arithmetic operation. The basic capabilities of a computer do not include this operation, and it must be programmed. However, routines exist for almost any computer and the problem-oriented languages con-

Table 1.3
Function Routines Assumed to Be Available

Type of Function	Name	Purpose
Sine	SIN(X)	Find sine of x
Cosine	COS(X)	Find cosine of x
Square root	SQRT(X)	Find square root of x
Natural logarithm (to the base e)	LOG(X)	Find logarithm of x to the base e
Exponential	EXP(X)	Find e^x
Arc tangent	ATAN(X)	Find arc tangent of x; $-\pi/2 <$ arctan $x < \pi/2$

sider it an arithmetic operation. Henceforth in this text it will be treated as one of the arithmetic operations of the computer and will be identified by a double star. A ** B will be interpreted to mean that the number in location A is raised to the power of the number in location B.

1.9 Summary

The digital computer is a machine that can perform only simple arithmetic and logic operations, input and output operations, and storage of the program and all required data. The storage unit has the characteristic of nondestructive read-out and destructive read-in. The program is a series of coded instructions that directs the computer on the computations and simple decisions to be made. The computer can execute a program written in machine language, which is a language of number codes. Machine-language programs, called compilers, can translate problem-oriented languages to machine language for execution by the machine.

References

1. Desmonde, W. H.: "Computers and Their Uses," Prentice-Hall, Inc., Englewood Cliffs, N.J., 1964.
2. Ledley, Robert S.: "Programming and Utilizing Digital Computers," McGraw-Hill Book Company, New York, 1962.
3. Sherman, Philip M.: "Programming and Coding Digital Computers," John Wiley & Sons, Inc., New York, 1963.

Chapter 2

The Flow Chart

2.1 Introduction Every problem prepared for machine computation must be organized into a sequence of arithmetic and logic statements that will lead to the desired solution. Provision must be made for placing in the memory of the machine all the data needed in the computations and for writing out the final results of the computations in a legible form. The vehicle for accomplishing these tasks is the *program.* It provides the control for all the operations to be carried out in the solution of the problem. In preparing a program an organizational chart is often made (or in simple cases merely visualized), showing the order in which the machine computations are to be performed, the logic steps to be taken, the information required as data, and the results to be written out. This chart shows how information flows through the computational unit of the machine and just what computations are to be made and can serve as a point of departure for writing an actual program for a machine. In fact, it is necessary only to set down appropriate coded instructions that accomplish the operations indicated on the chart to have a workable program. This chart is called the *flow chart* for the problem and represents the final objective for examples discussed in this text.

2.2 Flow-chart Symbols

Operations used for the flow chart are *input* and *output* statements, *arithmetic* statements, and *logic* statements. The input statement is denoted by READ and is used to introduce data into the machine. The arithmetic statement uses the operations *add, subtract, multiply,* and *divide.* The logic statement *tests the sign of a specified quantity and switches to different parts of the program depending on whether this quantity is negative, zero, or positive.* The machine can identify each alternative and chooses one of three paths on this basis. The output statement, denoted by WRITE, is used for the *display of results.* Little attention is paid in this book to limitations that may exist because of a particular machine, it being assumed that sufficient storage capacity exists for the program and data. The special symbolism used in the flow chart has been chosen for simplicity and is defined below.

The input statement is enclosed in a box with the upper left corner clipped. The word READ is followed by the list of terms to be introduced into the machine. The box in Fig. 2.1 indicates that data are to be transferred from the input unit into locations A, B, C, and D and that they are identified by these symbols in subsequent program statements.

The arithmetic statement is a substitution statement which says to do certain arithmetic operations with certain numbers and then store the result in a specified location in memory. Arithmetic statements are enclosed in rectangles. The statement in Fig. 2.2 indicates that the contents of A are to be added to the contents of B and the sum is to be stored in E. This statement defines the number in E. A and B must be defined and available to the machine

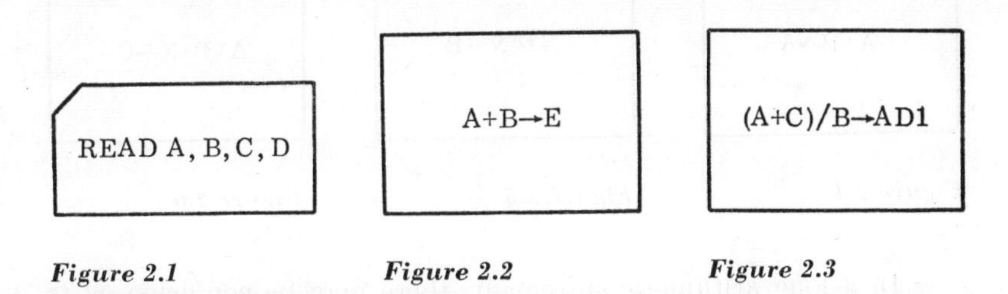

Figure 2.1 *Figure 2.2* *Figure 2.3*

before this statement is executed. The contents of locations A and B are not changed by the arithmetic statement. For convenience of notation, when arithmetic statements are used in the future, the arithmetic operation "the contents of A are added to the contents of B" will be denoted by A + B. The statement in Fig. 2.3 is defined to mean "add A to C, divide this sum by B, and then store the result in AD1." The slash is used to denote division. The statement in Fig. 2.4 is admissible and is defined to mean "add A to B, then store the result back in A." From the nondestructive read-out and destructive read-in property of the computer this statement *updates* A by B. After this statement is executed, location A contains the sum of B and the previous contents of A. The number originally contained in A is destroyed.

The symbols used to denote arithmetic operations are those common to almost all the problem-oriented languages. The addition and division symbols are identified above. The symbols for subtraction and multiplication are a minus and a star respectively. Thus

+	means addition
−	means subtraction
/	means division
*	means multiplication

The operation "raise to the power of," described in Chap. 1, is defined in Fig. 2.5. This statement means raise the number in X to the power of the number in N and store the result in B. After execution of this statement, B contains X raised to the power N.

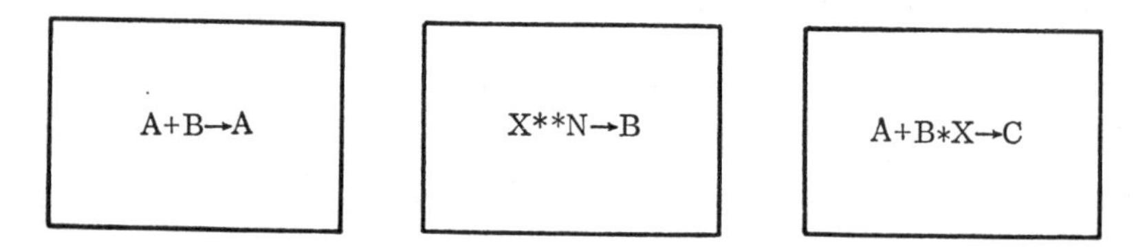

Figure 2.4 *Figure 2.5* *Figure 2.6*

In a long arithmetic statement, there may be confusion as to which operation should be performed first. For example, should the statement in Fig. 2.6 be interpreted as "compute the sum A + B, multiply this sum by X, then store the results in C" or as "compute the product B * X, add A to this product, then store the result in C"? To prevent this confusion, a definite hierarchy of operations is assumed, which agrees with conventional practice. First to be done is the raise-to-the-power-of operation, next the multiply and divide operations, and finally the add and subtract operations. This is the same hierarchy of operations used in algebra. When two operations of the same level in the hierarchy occur next to each other, as in Fig. 2.7, they proceed from left to right. The statement will be interpreted as "multiply A by B, divide this product by C, and store the result in D." This order can be changed by putting parentheses in the statement. The operations inside a set of parentheses are done first. For example, the box in Fig. 2.6 is interpreted to mean "compute the product of B and X, add A to this product, then store the result in C";

whereas, the statement in Fig. 2.8 is interpreted to mean "compute the sum of A and B, multiply this sum by X, then store the result in C," which differs from the result obtained in Fig. 2.6.

When there is more than one set of parentheses in an arithmetic statement, the operations proceed from the inner set of parentheses outward. The statement in Fig. 2.9 is defined to mean "multiply X by A, add B to the product, multiply the result by X, add C to the product, multiply the result by X, and add D to the product." The result is then stored in E.

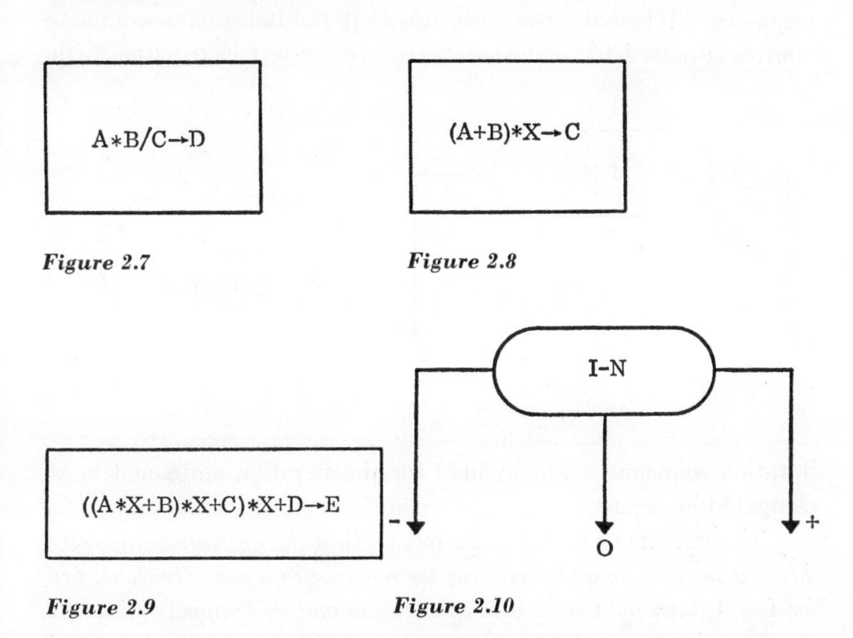

Figure 2.7

Figure 2.8

Figure 2.9

Figure 2.10

A round-ended rectangular figure signifies a logic statement. Figure 2.10 shows how the number obtained from the computation I − N is compared with zero. The argument of the logic statement may be any arithmetic expression. The value of the expression is evaluated using current values for the variables and one of the three possible branches taken, depending on the result. If the result is negative, the line marked "−" is followed; if the result is zero, the line marked "0" is followed; and if the result is positive, the line marked "+" is followed. Each line must be labeled to indicate the condition that is to be met before this branch is followed.

Two of the three branches can be combined if both lead to the same operation. If two branches are combined, two alternatives exist, and only two lines issue from the logic box. The combined branch is labeled with both symbols, as illustrated in Fig. 2.11.

The *iteration statement* signifies that a sequence of operations is to be repeated with some internal modifications until certain conditions are satisfied. This statement sets up the control of the computer to test for completion of the iteration and to perform certain indicated program modifications for each time through the sequence. When the test indicates that the iteration is complete, control is passed to the instruction after the last instruction in the

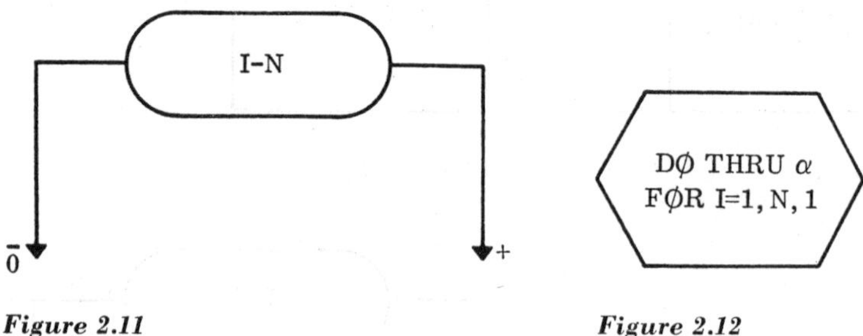

Figure 2.11 *Figure 2.12*

iteration sequence. The symbol for the iteration statement is an elongated hexagon.

The statement in Fig. 2.12 means that *the operations following this statement up to and including the one labeled* α *are to be done, first for* I = 1, *then for* I = 2, *and so on up to and including* I = N. In the flow chart α is the terminal point of the calculation loop, and it marks the end of the iteration. The sequence of calculations that comes under the control of the iteration statement is called the *scope* of the iteration. The iteration variable I is a dummy variable that is first set equal to 1, as indicated, and is then incremented by the value 1, which is the last number in the statement, for each iteration. This process continues until I is greater than N. The iteration is done for I = N but not for I > N.

Figure 2.13 illustrates the most general iteration statement used in this text, which in a flow chart would say: "Do the following operations up to and including β for J set equal to the contents of

K, then for J incremented by the contents of N, and so on, each time incrementing J by the contents of N, until the term J − L has the same sign as N." (This is to be interpreted that the operations are done for J = L.) The only restriction placed on these numbers is that N ≠ 0. If N is not explicitly given, it is assumed to be 1.

It is possible to include one or more iteration statements within the calculation loop of an iteration statement. This is called *nesting* (Fig. 2.14). The first statement sets up the control for iteration

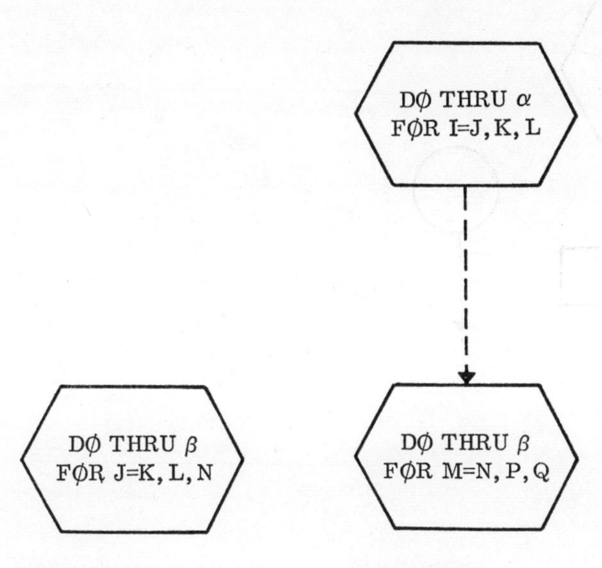

Figure 2.13 ***Figure 2.14***

on I and includes operations up to α in the flow chart, as described above. The next statement, which need not follow immediately after the first but must fit entirely within the scope of the first, sets up the control for an iteration on M and includes the operations up to β. β cannot appear after α in the routine, but it can come before α or it may be α. *The iterations on the inner loop must be completed before control can pass back to the outer, or first, loop.* In the first pass through the loop the first statement sets up the control to do the operations to α with I equal to the contents of J. When the second iteration statement is encountered, it sets up the control to do the operations to β first with M equal to the contents of N and then with M incremented by the contents of Q. The second iteration continues until it is completed without altering the outer loop. Control is then passed to the outer loop. On the second iteration through the outer loop I is updated by the contents of L. When the second statement is encountered, it assumes control, and the operations within its scope are completed. Control is then returned to the outer loop. Thus M takes on all its prescribed

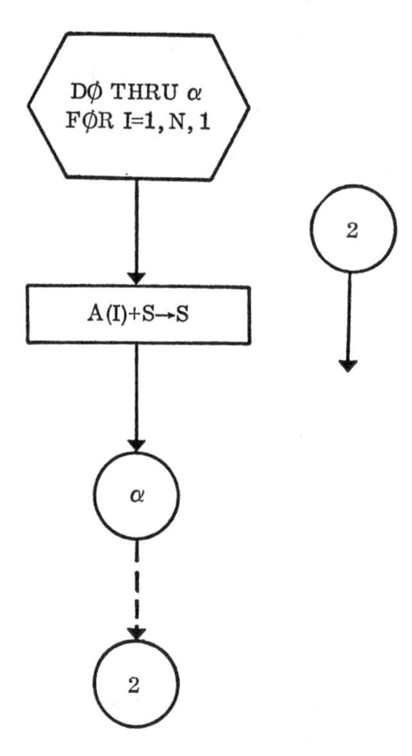

Figure 2.15

values, and the operations in the inner loop are repeated the specified number of times for each value of I in the outer loop. Use of the nested iteration statements is illustrated in Example 2.6.

A circle is used to define the end of an iteration loop (Fig. 2.15). Thus in the examples above, a circle containing α or β, whichever applies, indicates the end of the loop. The circle is also used to connect different parts of the chart that lie so far apart physically that a line cannot conveniently join them, in which case a circle containing the same symbol can be used to indicate that two parts of a chart should be connected (Fig. 2.15). Ordinarily lines are used to connect sequential instructions in the chart.

The output statement is symbolized in the flow chart by the sketch of a torn piece of paper. Figure 2.16 indicates that the contents at A, B, and C are to be transferred to the output unit. The contents of A, B, and C are not disturbed by this statement.

In all machines certain steps must be made to start the machine, and some signal to stop must be given. In the flow charts the beginning of the program is denoted by START and the end by STØP, both enclosed by a diamond. Every flow chart must have one start and at least one stop.

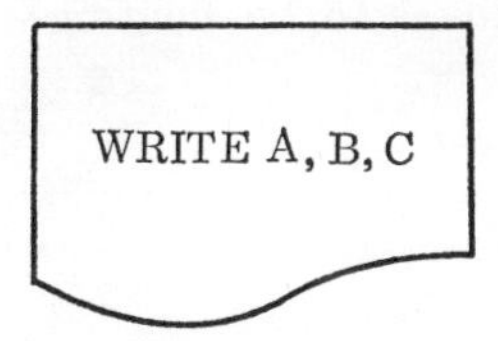

Figure 2.16

With this system of symbols, representing the capabilities of any modern computer, detailed organizational charts for machine computations can be made. When one knows a specific problem-oriented language and the read and write characteristics of a machine, it is easy to prepare a working program from a flow chart constructed along these lines.

2.3 *Variable Names*

The use of variable names to identify numerical data was illustrated in the preceding discussion on flow charts, but explicit rules must be observed if code names are to be kept in line with actual usage.

Variable names are usually limited in length because they must be coded as numbers and space is required to store these numbers in memory. The number of characters used to define a variable is arbitrary, provided that it does not exceed a given maximum, which is usually five or six, depending upon the language being used. In this book variable names will be limited to six characters. *The first character of any variable name must be alphabetic.* Additional characters may be alphabetic, numeric, or a combination of both. Permissible variable names are:

BUT5	A	BAC	IPUT
ALEXØ	ALU	SUM123	I11509

All alphabetic characters used in names are capitalized.

It was pointed out in Chap. 1 that numbers in the integer mode are coded differently from numbers in the floating-point mode. Some recognition of this fact is necessary in assigning variable names. In this book a list of variable names which represent integers is given on each flow chart. All variable names in the flow chart not appearing on the list are assumed to be floating-point variables.

2.4 Examples

A few sample problems will illustrate how a flow chart is constructed.

Example 2.1

Construct a flow chart that will find the sum of a sequence of numbers represented by $a_1, a_2, \ldots, a_n$.

This addition operation on the computer is done by finding first the partial sum of $a_1 + a_2$, then $a_1 + a_2 + a_3$, and so on, until each number has been added to the partial sum to obtain the total. Either of the two flow charts in Fig. 2.17 will accomplish the necessary operations, the two solutions being the same except for the control statements.

The first symbol in every flow chart is START. It indicates that appropriate control statements are to be placed ahead of the program for proper initiation of the machine. The READ statement means that the number n is read into a location labeled N and that the sequence of numbers $a_1, a_2, \ldots, a_n$ is read into the locations labeled A(1), A(2), . . . , A(N). The shorthand notation in this box will be used frequently. The number in location N indicates the number of items in the set. A(1), . . . , A(N) contain the numbers $a_1, \ldots, a_n$, respectively. The next statement is the first arithmetic statement and places a floating-point zero in the contents of a location called SUM.

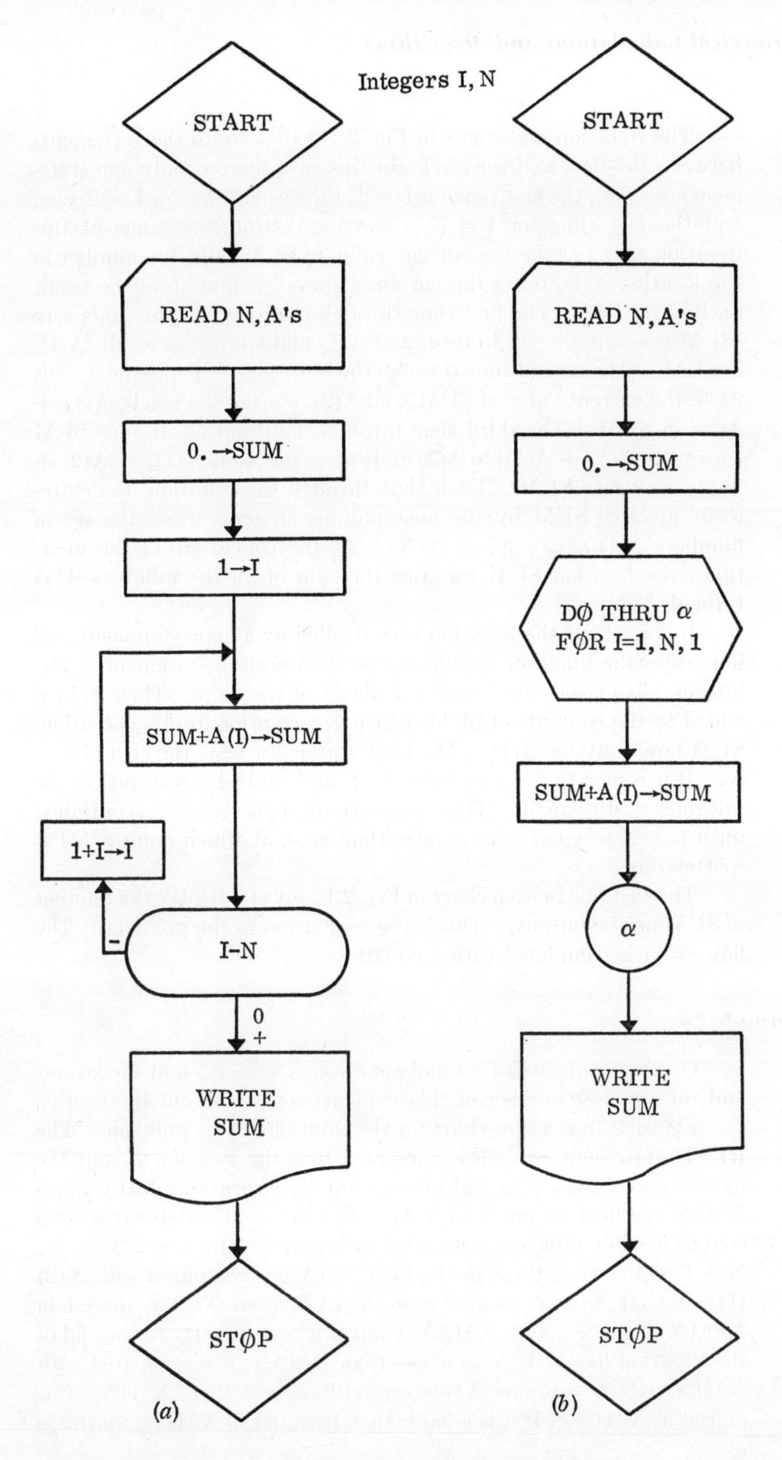

Figure 2.17 Compute the sum of a sequence of numbers.

The iteration statement in Fig. 2.17*b* says to do the statements between this box and the circle (in this case there is only one statement) N times, the first time for I = 1, the second time for I = 2, etc., and the last time for I = N. The box within the range of this iteration says to take the current value of SUM, add the number in the location A(I), using the current value of I, and store the result back into SUM. The first time through the iteration this adds zero (SUM was initially set to zero) and A(1) and stores the result, A(1), in SUM. The second time through the iteration, this statement adds A(1), the current value of SUM, and A(2), storing the result, A(1) + A(2), in SUM. The third time through the iteration, it adds SUM [currently A(1) + A(2)] to A(3) and stores the result, A(1) + A(2) + A(3), back into SUM. Each time through the iteration, the statement updates SUM by the next number in order from the set of numbers A(1), A(2), . . . , A(N). At the conclusion of the iterations, the location SUM contains the sum of all the numbers A(1) through A(N).

In Fig. 2.17*a* the iteration is controlled by a logic statement and illustrates the functions performed by the iteration statement. The integer 1 is placed into location I ahead of the loop. Then A(1) is added to the contents of SUM, which is zero prior to this operation. SUM now contains A(1). The logic statement tests the sign of I − N. If it is negative, I is updated by 1, and control is returned to the arithmetic statement. This sequence of operations is continued until I − N is equal to or greater than zero, at which point SUM is written out.

The last box in each chart in Fig. 2.17 says to display the number in SUM on the output. This is the conclusion to the problem. The flow chart is completed with a STØP.

Example 2.2

Given a sequence of n numbers $x_1, x_2, \ldots, x_n$, find the largest and the smallest number of this sequence and write out the results.

Figure 2.18 is a flow chart for the solution of this problem. The READ statement reads the number n into the location N and the numbers $x_1, x_2, \ldots, x_n$ into the vector X. Locations XMAX and XMIN are next set equal to X(1). The iteration statement sets up control for doing the operations that follow up to α for I = 2, 3, . . . , N. The first time through the loop XMAX is compared with X(2) (I = 2). If X(2) is greater than XMAX, then X(2) is placed in XMAX; if it is equal to XMAX, control is sent directly to the end of the iteration loop. If X(2) is less than XMAX, it is compared with XMIN, which contains X(1). If X(2) is less than XMIN, it is placed in XMIN. If it is greater than or equal to XMIN, control is

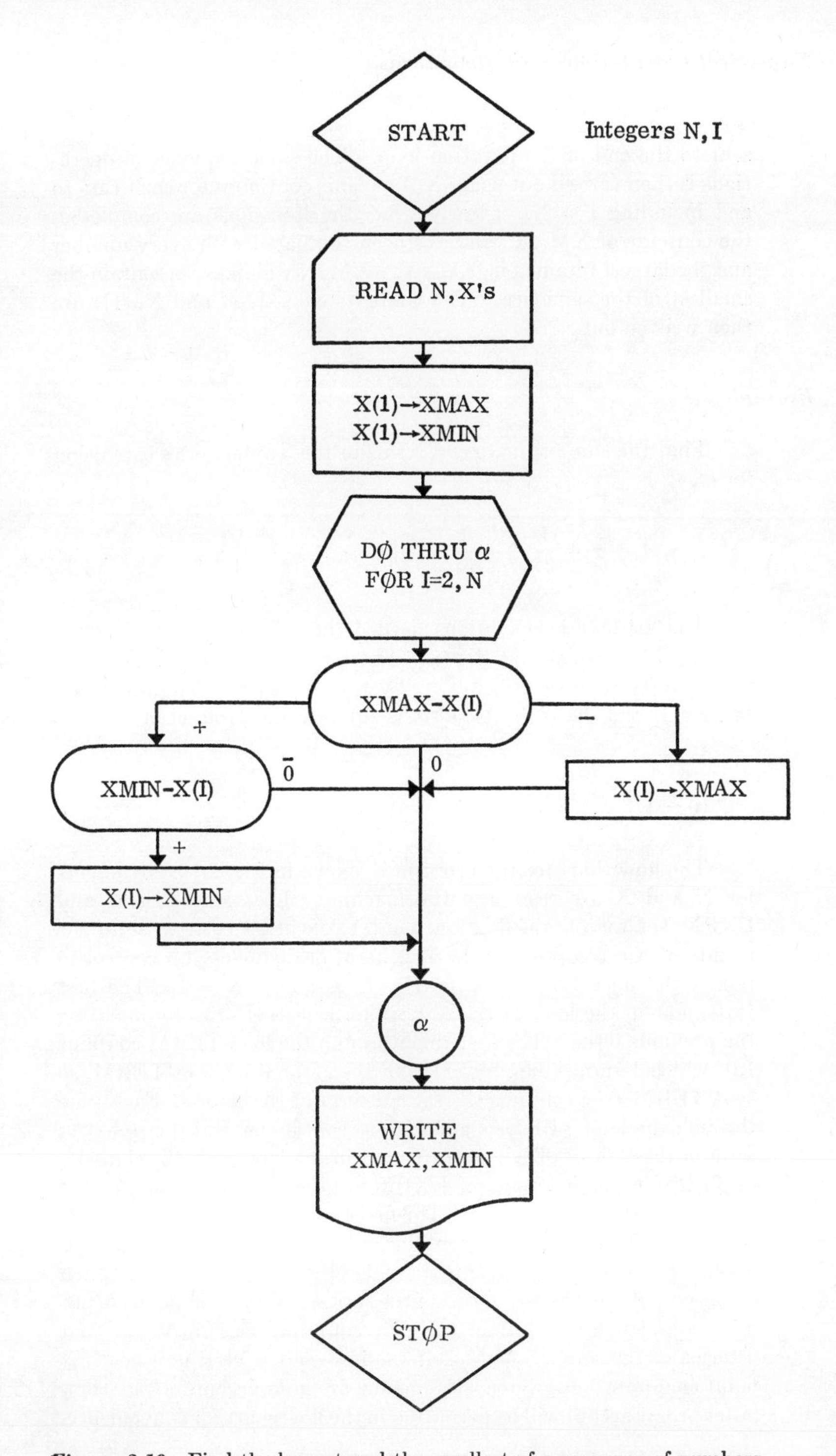

Figure 2.18 Find the largest and the smallest of a sequence of numbers.

sent to the end of the iteration loop. The same sequence of operations is then carried out using X(3). This continues for each I up to and including I = N. Clearly when the iterations are completed, the contents of XMAX will have been compared with every number and the largest retained in XMAX. XMIN will likewise contain the smallest of the sequence. The contents of XMAX and XMIN are then written out.

Example 2.3

Find the sum of the first n terms in the Taylor-series expansion for e^x.

$$e^x = 1 + x + \frac{1}{2!}x^2 + \frac{1}{3!}x^3 + \cdots + \frac{x^{n-1}}{(n-1)!}$$

The arithmetic operations needed for evaluating this sum are simplified by noting that the $(i + 1)$st term in the series is obtained from the ith term by multiplying by x/i. The second term is equal to $1 \cdot x/1$, and the $(i + 1)$st term is equal to the product of

$$\frac{x^{i-1}}{(i-1)!} \qquad \text{and} \qquad \frac{x}{i}$$

The flow chart for the problem is shown in Fig. 2.19. Numbers for N and X are read into the machine. Locations TERM and EXPX, set to contain floating point 1.0, will be used to store the results of the computations. The iteration loop sets up control to repeat the operations to α for I = 1, 2, . . . , N − 1. The first statement in the loop gives the next term in the series by updating the previous term. The first time through the loop TERM contains 1.0, which is multiplied by $x/1$, and then stored back in TERM, so that TERM now contains the second term of the series. Each time through the loop TERM is updated, so that it contains the $(i + 1)$st term in the series. The last statement in the loop adds the contents of TERM to the contents of EXPX to give the partial sum of the first $i + 1$ terms of the series. The first time through the loop EXPX contains 1.0, and TERM contains $x/1.0$ when this statement is executed. The updated contents of EXPX are $1.0 + x/1.0$, which is the sum of the first two terms of the series. When the iteration has been completed $n - 1$ times, EXPX contains the sum of the first n terms of the series. Note that the integer I is used in a floating-point computation. Such statements are not acceptable in computer programs but will be permitted in the flow chart for convenience.

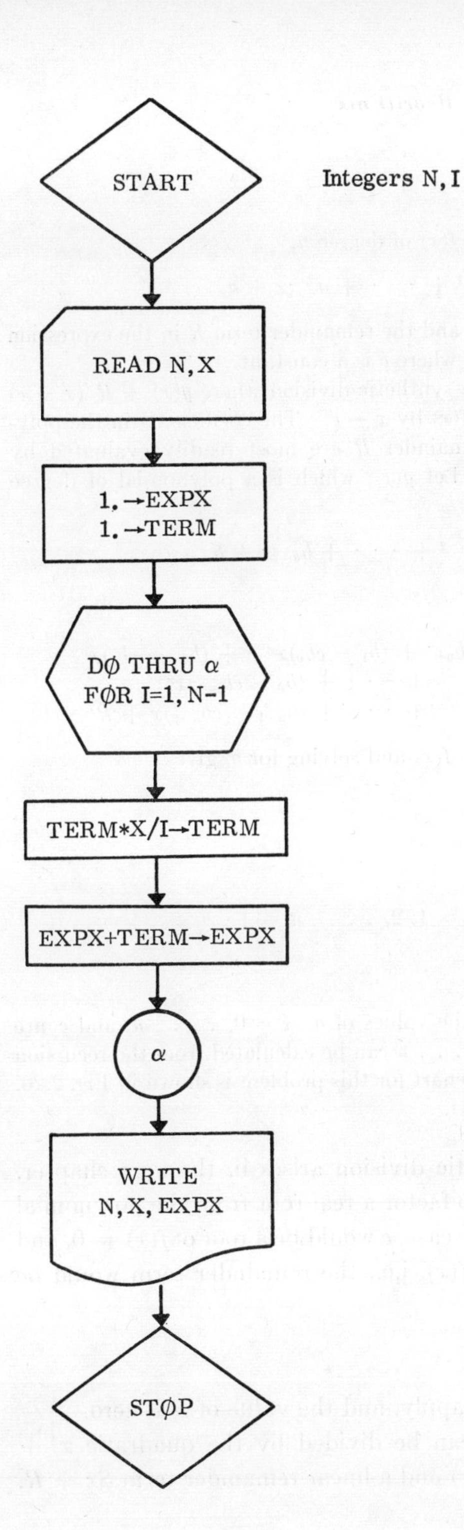

Figure 2.19 Compute the first n terms in the series for e^x.

Example 2.4

Given a polynomial $f(x)$ of degree n,

$$f(x) = a_0x^n + a_1x^{n-1} + \cdots + a_{n-1}x + a_n$$

find the polynomial $g(x)$ and the remainder term R in the expression $f(x) = (x - c)g(x) + R$, where c is a constant.

This is a problem in synthetic division where $g(x) + R/(x - c)$ is the result of dividing $f(x)$ by $x - c$. The coefficients in the polynomial $g(x)$ and the remainder R are most readily evaluated by comparing coefficients. Let $g(x)$, which is a polynomial of degree $n - 1$, be represented by

$$g(x) = b_0x^{n-1} + b_1x^{n-2} + \cdots + b_{n-2}x + b_{n-1}$$

Then

$$\begin{aligned} R + (x - c)g(x) = b_0x^n &+ (b_1 - cb_0)x^{n-1} + (b_2 - cb_1)x^{n-2} \\ &+ \cdots + (b_i - cb_{i-1})x^{n-i} \\ &+ \cdots + (b_{n-1} - cb_{n-2})x + R - cb_{n-1} \end{aligned}$$

Comparing coefficients in $f(x)$ and solving for b_i gives

$$b_0 = a_0$$

$$b_1 = a_1 + cb_0$$

$$b_i = a_i + cb_{i-1} \qquad i = 1, 2, \ldots, n - 1$$

$$b_n = a_n + cb_{n-1}$$

where $b_n = R$. When the values of a_i, $i = 0, \ldots, n$, and c are known, each b_i, $i = 0, \ldots, n$ can be calculated from the recursion formulas above. A flow chart for this problem is shown in Fig. 2.20. R will be in location B(N).

The problem of synthetic division arises in the next chapter, where it may be necessary to factor a real root from the polynomial $f(x) = 0$. Should this be the case, c would be a root of $f(x) = 0$, and $x - c$ would be a factor of $f(x)$, i.e., the remainder term would be zero.

$$f(x) = (x - c)g(x)$$

The results for Example 2.4 apply, and the value of R is zero.

In a similar way $f(x)$ can be divided by the quadratic $x^2 + ux + v$ to give a quotient $g(x)$ and a linear remainder term $Sx + R$.

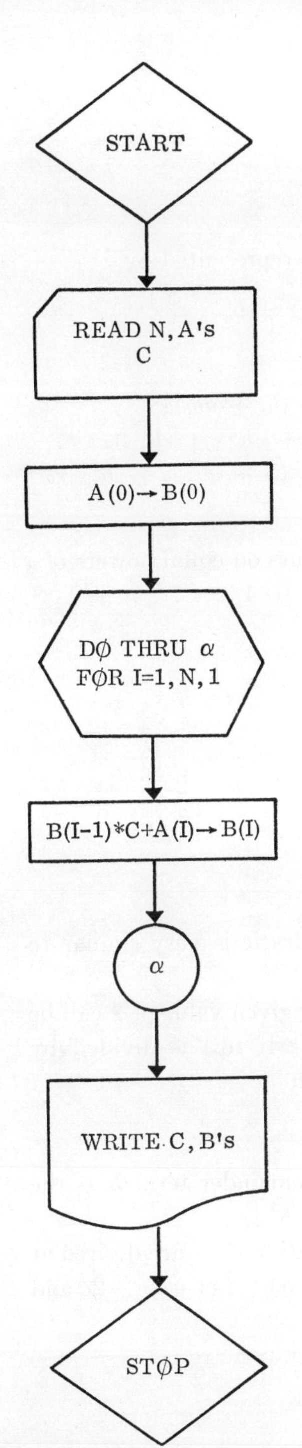

Figure 2.20 Synthetic division.

This is indicated by

$$f(x) = (x^2 + ux + v)g(x) + Sx + R$$

Let $g(x)$, a polynomial of degree $n - 2$, be represented by

$$g(x) = b_0x^{n-2} + b_1x^{n-3} + \cdots + b_{n-3}x + b_{n-2}$$

Then

$$\begin{aligned} &Sx + R + (x^2 + ux + v)g(x) = b_0x + (b_1 + ub_0)x^{n-1} \\ &+ (b_2 + ub_1 + vb_0)x^{n-2} + \cdots + (b_i + ub_{i-1} + vb_{i-2})x^{n-i} \\ &+ \cdots + (b_{n-2} + ub_{n-3} + vb_{n-4})x^2 + (S + ub_{n-2} + vb_{n-3})x \\ &+ (R + vb_{n-2}) \end{aligned}$$

Comparing with $f(x)$ and equating coefficients on equal powers of x gives the following expression for b_i, $i = 0, 1, \ldots, n-2$, S, and R:

$$b_0 = a_0$$

$$b_1 = a_1 - ub_0$$

$$b_i = a_i - ub_{i-1} - vb_{i-2} \qquad i = 2, 3, \ldots, n-2$$

$$S = b_{n-1} = a_{n-1} - ub_{n-2} - vb_{n-3}$$

$$R = a_n - vb_{n-2}$$

The flow chart for dividing $f(x)$ by a quadratic is very similar to that in Fig. 2.20 and is not given here.

The evaluation of a polynomial for any given value of x can be done by the above computation routines. If $f(x)$ is divided by $x - c$ to get $g(x)$ and the remainder R, as in

$$f(x) = (x - c)g(x) + R$$

then clearly $f(c) = R$. The value of the remainder term R is the value of $f(x)$ at $x = c$.

If the values of both $f(x)$ and its derivative $f'(x)$ are desired at $x = c$, the division by a quadratic can be used. Let $u = -2c$ and $v = c^2$; then

$$\begin{aligned} f(x) &= (x^2 + ux + v)g(x) + Sx + R \\ &= (x^2 - 2cx + c^2)g(x) + Sx + R \\ &= (x - c)^2g(x) + Sx + R \end{aligned}$$

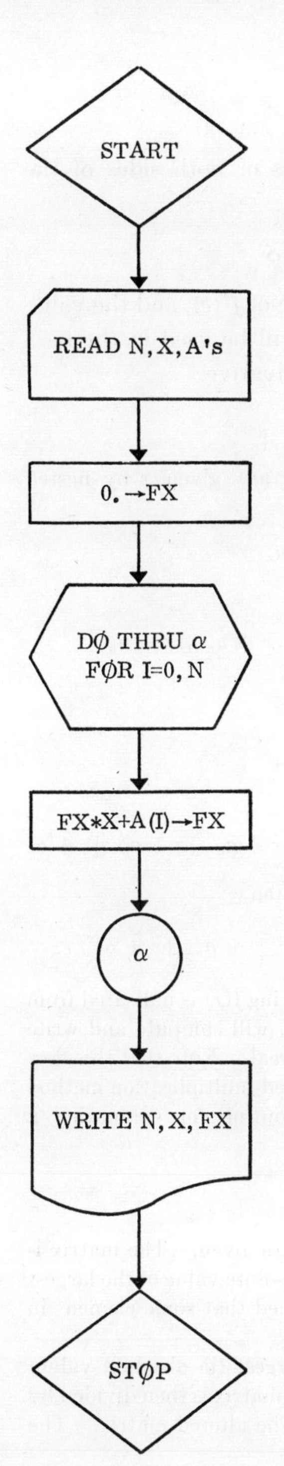

Figure 2.21 Evaluate a polynomial by nested multiplication.

Clearly, $f(c) = Sc + R$. Taking derivatives of both sides of the equation with respect to x gives

$$f'(x) = 2(x - c)g(x) + (x - c)^2g'(x) + S$$

Then $f'(c) = S$. The value of S is the value of $f'(c)$, and the value of $Sc + R$ is the value of $f(c)$. This fact will be used in the next chapter to evaluate a polynomial and its derivative.

Example 2.5

Find the numerical value of $f(x)$ for any given x by nested multiplication.

$$f(x) = a_0x^n + a_1x^{n-1} + \cdots a_{n-1}x + a_n$$

Let x be factored out of the first n terms:

$$f(x) = (a_0x^{n-1} + a_1x^{n-2} + \cdots + a_{n-2}x + a_{n-1})x + a_n$$

x is now factored from the expression

$$a_0x^{n-1} + \cdots + a_{n-2}x$$

to give

$$f(x) = [(a_0x^{n-2} + a_1x^{n-3} + \cdots + a_{n-3}x + a_{n-2})x + a_{n-1}]x + a_n$$

This procedure is repeated until $f(x)$ is written

$$f(x) = \{\cdots [(a_0x + a_1)x + a_2]x + \cdots + a_{n-1}\}x + a_n$$

A simple calculation routine for computing $f(x)$ is indicated from this expression. The flow chart (Fig. 2.21) will compute and write out the value of $f(x)$ for any given x that is real. Note that this flow chart is very similar to Fig. 2.20. The nested-multiplication method is equivalent to the synthetic-division routine for evaluating a polynomial.

Example 2.6

The elements of an n by m matrix A are given. The matrix is altered by dividing each element by the absolute value of the largest numerical entry in the matrix. It is assumed that some elemen in the matrix is different from zero.

In the solution of this problem the largest (in absolute value) element of A is found. Each element of the matrix is then divided by the absolute value of this entry to obtain the altered matrix. The flow chart is given in Fig. 2.22.

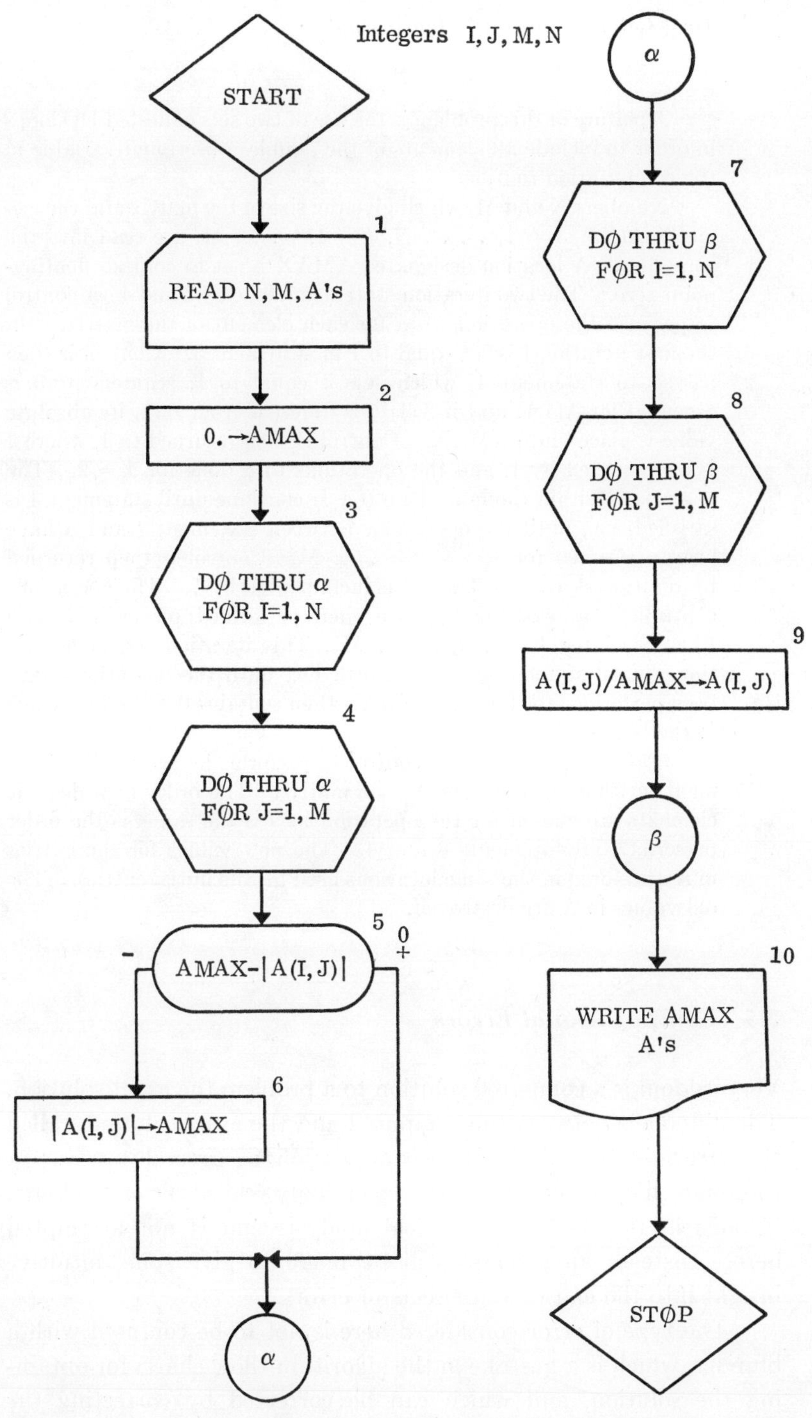

Figure 2.22 Divide each number in an array by the largest number in the array.

A feature of this problem is the use of two sets of nested DØ loops in order to include all elements of the double-subscripted variable in the computation routine.

Numbers N and M, which give the size of the matrix and the elements A(I,J), I = 1, . . . , N, J = 1, . . . , M, are read into the computer. A location designated AMAX is set to contain floating-point zero. The two iteration statements labeled 3 and 4 set control to perform the operations to α for each element of the matrix. On the first iteration I is set equal to 1 in statement 3. Control is then passed to statement 4, which sets J equal to 1. Statement 5 is executed for A(1,1) and if A(1,1) is different from zero, its absolute value is placed into AMAX. Control is then returned to 4, where J is incremented by 1, and the operations to α done for J = 2. The operations within the inner loop (on J) continue until statement 4 is satisfied, i.e., until the operations between statement 4 and α have been performed for J = 1, 2, . . . , M. Control is then returned to iteration statement 3, and I is incremented by 1. The operations controlled by statement 4 are then performed for each J with I = 2, i.e., for J = 1, 2, . . . , M. This iteration sequence continues until statement 3 is satisfied, i.e., until the operations have been performed for I = N. AMAX then contains the absolute value of the largest element of A.

Statements 7 and 8 set control to perform the arithmetic statement in 9 for each element of the matrix. The order in which the elements are chosen for the operation in 9 is the same as the order prescribed in statements 3 and 4. The new values for the entries in A are stored in the same locations used for the initial entries. The old values in A are destroyed.

2.5 Computational Errors

Very seldom is a computed solution to a problem the exact solution. The difference between the computed and the exact solution, called the error, is the result of a combination of several causes. A thorough discussion of errors, commonly called *error analysis*, belongs in the field of numerical analysis and is not attempted here. Instead, an attempt will be made to give some intuitive insight into the causes and effects of errors.

The type of error considered here is not to be confused with a blunder, which is a mistake in the algorithm (flow chart) for obtaining the solution, and which can be corrected by correcting the algorithm. An error is usually a consequence of the method or machine used and is generally unavoidable. For example, no matter how hard one tries, one cannot measure the length of this page exactly. The ruler is imprecise to some degree, and the graduations on it are only so close together. No matter how accurately it is measured, the figure is still only approximate, and the error, albeit very small, still exists. If the length of this page is an input number for a program, an unavoidable error has been introduced with the input data. Thus, one source of error is the error in the input data.

A second source of error was illustrated by Example 2.3, which was to compute e^x using the first n terms in the Taylor-series expansion. The number computed was not e^x but $y(x)$, where

$$y(x) = 1 + x + \frac{1}{2!}x^2 + \frac{1}{3!}x^3 + \cdots + \frac{x^{n-1}}{(n-1)!}$$

The error $e^x - y(x)$ caused by truncating the infinite series for e^x after n terms is called *truncation error*. Since using all the terms in an infinite series would require an infinite amount of computer time and could not possibly be accomplished, truncation error is unavoidable. However, it can be made as small as desired simply by taking more terms in the series.

Often the size of the truncation error depends upon a parameter h. An important bit of information is how the error depends on h. It is usually written as $O(h^n)$, read "of order h^n." If the truncation error is $O(h^3)$, and if the computations are performed twice, once with $h = k$ and once with $h = k/2$, the truncation error for the second computation will be approximately $\frac{1}{8} = (\frac{1}{2})^3$ the error in the first computation. The $O(h^n)$ notation will be used in this text whenever appropriate to indicate truncation error.

A third source of error is the way numbers are manipulated in the computer. They are generally handled as scientific numbers (see Sec. 1.5) with a fixed number of significant digits. For example, if a four-decimal digit machine is used to compute e^{-10} in Example 2.3, the value for TERM (see the flow chart in Fig. 2.19) reaches a maximum of 2,756, and the sum reaches a value of $-1{,}414$ in the course of the computations. If the iterations are continued until the contribution of the number in TERM does not change the sum, a value of -0.2015 is obtained. This result of -0.2015 is far from the true value of $e^{-10} = 0.0000454$. The error caused by rounding off the numbers to four decimal digits is called *round-off error*. The size of the round-off error depends upon the particular machine used. It can be reduced only by carrying more digits in the computations.

These three sources of error all contribute to the fourth, the propagation of errors through the computations. The assumption that the effect of a very small error introduced early in the computations will remain small, i.e., that the numerical calculations are stable, is not always valid. If the magnitude of an error is multiplied by 1.01 for each computation, then, after 100 computations the magnitude will be 2.7 times the original magnitude. While this may be acceptable, the magnitude of the error would be 21,000 times as great after 1,000 computations and 440,000,000 times as great, probably invalidating any results, after 2,000 computations. This emphasizes the importance of stability when many computations are to be made. The problem of stability is very difficult because it depends not only on the numerical method being used but also on the problem being solved.

Any computed number is affected by some combination of these sources of error. In most problems the input data will be approximate, truncated formulas will be used, the numbers will be rounded to a given number of significant digits, and the numerical procedure, while stable at some point in the computations, may be unstable at another point. Just because a machine-computed answer has 16 digits it should never be interpreted as being correct to this number of places. In fact, it may not be correct to even the first digit. *All computed results should be checked for validity by any means available.*

2.6 Summary

The flow chart is a pictorial presentation of the calculation procedures used in solving a problem. It shows the calculations that are made, the order in which they are made, and the logic used in the problem. Actual computer programs can be transcribed with little change from a good detailed flow chart as outlined in this chapter. Several examples illustrate the use of the flow chart for outlining the solution of a problem. Errors arising in numerical computations are discussed.

Problems

1. Write a flow chart that will place the numbers in a_i, $i = 1, 2, \ldots, n$, that are greater than x into the vector B.

2. Write a flow chart to arrange the numbers a_i, $i = 0, 1, 2, \ldots, n$, in order of magnitude, $a_{i+1} \leq a_i$, $i = 0, 1, 2, \ldots, n - 1$.

3. Write a flow chart to arrange the numbers a_i, $i = 0, 1, 2, \ldots, n$, in order of absolute value.

4. Write a flow chart to find the product $\prod_{i=1}^{n} i = 1 \cdot 2 \cdot 3 \cdot \cdots \cdot n$.

5. Write a flow chart to find the sum $\sum_{i=1}^{n} i$.

6. Write a flow chart to find the scalar product of the vectors A and B. $AB = \sum_{i=1}^{n} a_i b_i$.

7. Do Prob. 6 for m sets of vectors A and B.

8. Write a flow chart to solve the quadratic equation $ax^2 + bx + c = 0$. Admit the possibility of complex roots and their identification.

9. Write a flow chart to find the sum $\sum_{i=1}^{n} i!$.

10. Write a flow chart to compute the product $\prod_{i=0}^{n} x_i$.

11. Write a flow chart to find the sum $\sum_{i=0}^{n} a_i + jb_i$, where $j = \sqrt{-1}$. Store the real part in RE and the imaginary part in IE.

12. Write a flow chart to find the product $\prod_{i=0}^{n} a_i + jb_i$, $j = \sqrt{-1}$.

13. Write a flow chart to perform the operation $\left(\sum_{i=0}^{n} a_i + jb_i\right) / (c + jd)$.

14. Write a flow chart to perform the division $\left(\sum_{i=0}^{n} a_i x^i\right) / (ax + b)$.

15. Write a flow chart to find the amount of money paid each month to pay off a loan of $10,000 in 20 years if the interest is 5 percent and is compounded monthly.

16. Write a flow chart to solve Prob. 15 if the interest is compounded daily.

17. Write a flow chart to find the sum of the first n terms of the sine series.

18. Do Prob. 17 for the cosine series.

19. Write a flow chart to multiply each element of the matrix $A = [a_{i,j}]$, $i = 0, 1, \ldots, n$, $j = 0, 1, 2, \ldots, m$, by the constant c. Store the elements back in the same locations.

20. Write a flow chart to place the elements of a matrix $A = [a_{i,j}]$, $i = 1, \ldots, n$, $j = 1, \ldots, m$, into the vector S according to the formula $a_{i,j} = s_{(i-1)n+j}$.

21. Write a flow chart to reverse the transformation in Prob. 20, i.e., $s_{(i-1)n+j} \rightarrow a_{i,j}$.

22. Write a flow chart that will change the sign on all negative elements in the n by m matrix A.

23. Write a flow chart to find the maximum of the sum of each of m columns in an n by m matrix.

24. Find the sum of the first n terms of the geometric series $\sum_{i=1}^{n} \frac{1}{i}$.

25. Compute the area of n triangles if two sides b and c and the angle θ between them are given. Area $= \frac{1}{2}bc \sin \theta$.

26. Write a flow chart to find the sum of the first n terms in the following series:

(a) $\log_e (1 + x) = x - \frac{1}{2}x^2 + \frac{1}{3}x^3 - \frac{1}{4}x^4 + \cdots$

(b) $(1 + x)^p = 1 + px + \frac{p(p-1)}{2!} x^2 + \frac{p(p-1)(p-2)}{3!} x^3 + \cdots$

(c) $\text{Arcsin } x = x + \frac{1}{2}\frac{x^3}{3} + \frac{1}{2}\frac{3}{4}\frac{x^5}{5} + \frac{1}{2}\frac{3}{4}\frac{5}{6}\frac{x^7}{7} + \cdots$

(d) $\text{Arctan } x = x - \frac{1}{3}x^3 + \frac{1}{5}x^5 - \frac{1}{7}x^7 + \cdots$

(e) $\log_e x = 2\left(y + \frac{1}{3}y^3 + \frac{1}{5}y^5 + \frac{1}{7}y^7 + \cdots\right)$, where $y = \frac{x-1}{x+1}$

(f) $\sinh x = x + \frac{1}{3!}x^3 + \frac{1}{5!}x^5 + \frac{1}{7!}x^7 + \cdots$

(g) $\cosh x = 1 + \frac{1}{2!}x^2 + \frac{1}{4!}x^4 + \frac{1}{6!}x^6 + \cdots$

(h) $\text{Arcsinh } x = x - \frac{1}{2}\frac{x^3}{3} + \frac{1}{2}\frac{3}{4}\frac{x^5}{5} - \frac{1}{2}\frac{3}{4}\frac{5}{6}\frac{x^7}{7} + \cdots$

(i) $\text{Arctanh } x = x + \frac{1}{3}x^3 + \frac{1}{5}x^5 + \frac{1}{7}x^7 + \cdots$

27. Write a flow chart to compute $\sqrt{x}$ using the following approximate formula. Test the result.

$$\sqrt{x} \doteq \frac{1 + 28x + 70x^2 + 28x^3 + x^4}{8(1 + 7x + 7x^2 + x^3)}$$

28. Find the coefficients in $g(x)$ and the remainder r, where $f(x) = (x - c)g(x) + r$, for

$$f(x) = x^4 - 28x^3 + 70x^2 - 28x + 1$$

when

(a) $c = 0$ (b) $c = 1$
(c) $c = -1$ (d) $c = 2$

29. Evaluate $f(x)$ and $f'(x)$ at $x = 0, 1, 2, 3$, where

$$f(x) = x^6 + x^5 - x^4 + x^3 - x^2 + x - 1$$

30. Write a flow chart to evaluate $f(x)$ and $f'(x)$, where $f(x)$ is an nth-degree polynomial.

References

1. Colman, Harry L., and Clarence P. Smallwood: "Computer Language: An Autoinstructional Introduction to Fortran," McGraw-Hill Book Company, New York, 1962.
2. Ledley, Robert S.: "Programming and Utilizing Digital Computers," McGraw-Hill Book Company, New York, 1962.
3. McCracken, Daniel D.: "A Guide to Fortran Programming," John Wiley & Sons, Inc., New York, 1961.
4. ———: "A Guide to ALGOL Programming," John Wiley & Sons, Inc., New York, 1962.
5. Organick, E. I.: "A Computer Primer for the MAD Language," The University of Michigan Press, Ann Arbor, Mich., 1961.
6. Plumb, S. C.: "Introduction to Fortran," McGraw-Hill Book Company, New York, 1964.
7. Sherman, Philip M.: "Programming and Coding Digital Computers," John Wiley & Sons, Inc., New York, 1963.
8. Wilkinson, James H.: "Rounding Errors in Algebraic Processes," Prentice-Hall, Inc., Englewood Cliffs, N.J., 1964

Chapter 3
Nonlinear Algebraic Equations

3.1 Introduction A nonlinear algebraic equation in the variable z is represented by

$$f(z) = 0 \tag{3.1}$$

In this chapter attention is directed primarily to the nth-degree polynomial in z.

$$f(z) = a_0z^n + a_1z^{n-1} + \cdots a_{n-1}z + a_n = 0 \tag{3.2}$$

The coefficients a_i, $i = 0, 1, \ldots, n$, are constants, which may be real or complex. Transcendental equations in z which include trigonometric and exponential forms such as

$$z \tan z - \cosh z = 0 \tag{3.3}$$

will also be considered.

A solution of Eq. (3.1) is a value of z which makes $f(z)$ equal to zero. In general there are n values of z which satisfy the nth-degree polynomial in Eq. (3.2), and there is an infinite number of values of z which satisfy the transcendental equation (3.3). The values of z which satisfy Eq. (3.1) are called *roots* of the equation.

Nonlinear algebraic equations arise in problems of finding natural frequencies of multi-degree-of-freedom systems, stability loads of structures, and oscillating frequencies of electrical circuits, to name a few examples. In physical problems leading to equations of the type in (3.1) the equation is often called the *characteristic equation*, and the roots are the *characteristic values* of the problem.

Direct methods are available for solving a quadratic and a cubic for its roots. Higher-degree polynomials and the transcendental equations must generally be solved by some approximate method. Those methods which have proved most satisfactory for high-speed computers are iterative schemes based upon successive improvement of some initial approximation to a root. All the methods presented in this chapter are of this type. The method of simple iteration is presented because of its historical importance and because it serves to illustrate in a simple way the basic iterative procedure. Those methods having the greatest utility for solving equations on automatic computers are the interval-halving, secant, Newton-Raphson, Lin-Bairstow, and Müller methods. Each is presented in some detail.

3.2 Method of Simple Iteration

The method of simple iteration is presented here primarily to illustrate the general iterative procedure used for solving algebraic equations of the type in Eq. (3.1). It is not generally recommended for use on an automatic computer, although it may be used in some cases with acceptable results. The algorithm for the simple iteration method is obtained by writing Eq. (3.1) as

$$f(z) = z - F(z) = 0 \tag{3.4}$$

and solving for z,

$$z = F(z) \tag{3.5}$$

Equation (3.5) is the iterative formula for improving an initial approximation to the root. If $z = z_0$ is the initial approximation, z_0 is placed in the right-hand side of Eq. (3.5) to give the first value in the iteration. Let this value be z_1.

$$z_1 = F(z_0) \tag{3.6}$$

The function $F(z)$ is then evaluated at $z = z_1$ to give the second iterate. The process is continued according to

$$z_{k+1} = F(z_k) \tag{3.7}$$

until a satisfactory approximation is made or until it is established that the process is not converging to a root.

A graphical interpretation of the simple iteration method is given in Fig. 3.1, in which the functions $y = z$ and $y = F(z)$ are plotted. In Fig. 3.1*b* $F(z)$ has a much steeper slope than in Fig. 3.1*a*. The intersection of the two curves in each figure defines a value of z which is equal to $F(z)$ and is therefore a root of $f(z) = 0$. The starting value for the iteration is z_0, shown in each graph. The value of $F(z_0)$, given by the point on the curve $F(z)$ at $z = z_0$, is the value of z_1 [see Eq. (3.6)]. At $z = z_1$, obtained graphically by projecting onto the line $y = z$, the value of $F(z_1)$ is obtained. This gives z_2, as shown, and the process may be continued for additional values in the sequence.

The behavior of the procedure in Fig. 3.1*a* and *b* is entirely different. In case *a* the sequence $z_1, z_2, \ldots$ converges to the point of intersection, whereas in case *b* it diverges and therefore cannot be used to find the root. The critical factor in the behavior of the method is the slope of the function $F(z)$ in the vicinity of the intersection. If the slope of $F(z)$ is less than 1 in absolute value, the process will converge to the root. It will be quite slow if $|F'(z)|$ is only slightly less than 1. If $|F'(z)| > 1$, the sequence will diverge, as illustrated in Fig. 3.1*b*. The simple iteration method is not

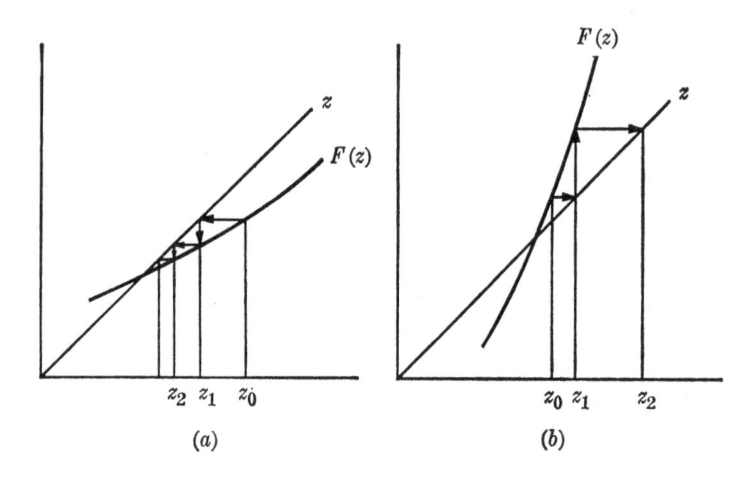

Figure 3.1 Simple iteration.

generally recommended for use on automatic computers. Under favorable conditions of convergence, however, it can give an acceptable and easily programmed method for estimating a root.

3.3 Interval Halving

Real roots of Eq. (3.1) are values of z for which the function $f(z)$ crosses the z axis, as illustrated in Fig. 3.2. The method of interval halving is based on finding an interval within which the curve crosses the axis and then repeatedly dividing by 2 the interval on which the intersection occurs.

In Fig. 3.2 the curve crosses the axis between z_1 and z_2 and has a real root at α on the interval of length Δz. The first approximation to the root in the interval-halving method z_3 is obtained by halving the interval and adding this increment to z_1:

$$z_3 = z_1 + \frac{\Delta z}{2} = \frac{z_1 + z_2}{2} \tag{3.8}$$

The root α will lie in the interval z_1 to z_3 or z_3 to z_2, provided, of course, that z_3 is not precisely equal to α. The interval on which the root is located is determined by comparing the signs of $f(z)$ at $z = z_1$ and $z = z_3$. If $f(z_1)$ has the same sign as $f(z_3)$, they are on the same side of the axis, and the root is between z_3 and z_2; if the signs are different, the root is between z_1 and z_3. In the example in Fig. 3.2 the root is between z_3 and z_2, and $f(z_1)$ and $f(z_3)$ have the same sign.

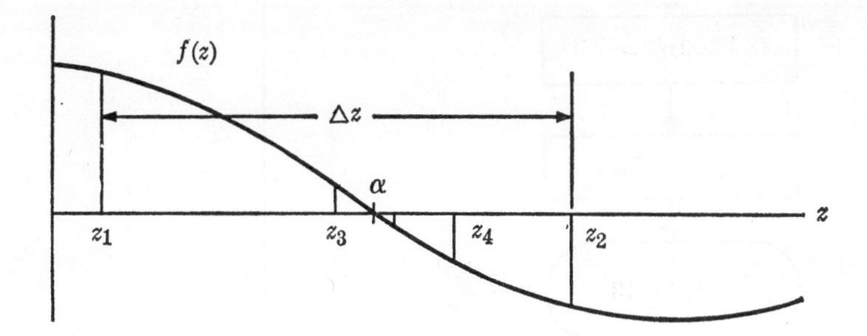

Figure 3.2 Interval-halving method.

The next approximation is obtained by dividing the reduced interval to obtain z_4. The value of the function is then computed at z_4 to determine the interval on which the root is located. This procedure is continued until the length of the divided interval is less than the admissible error in the root.

A flow chart for the interval-halving method is given in Fig. 3.3. An explanation of the flow chart follows.

1. The interval on which the root is located, designated by z_1 and z_2, is read into locations Z1 and Z2. ERR contains the admissible error in the root.

2. The value of the function at z_1 is computed and placed in location FZ1. This is a symbolic notation that denotes the necessary routine for finding $f(z)$. This notation will be used extensively in this text.

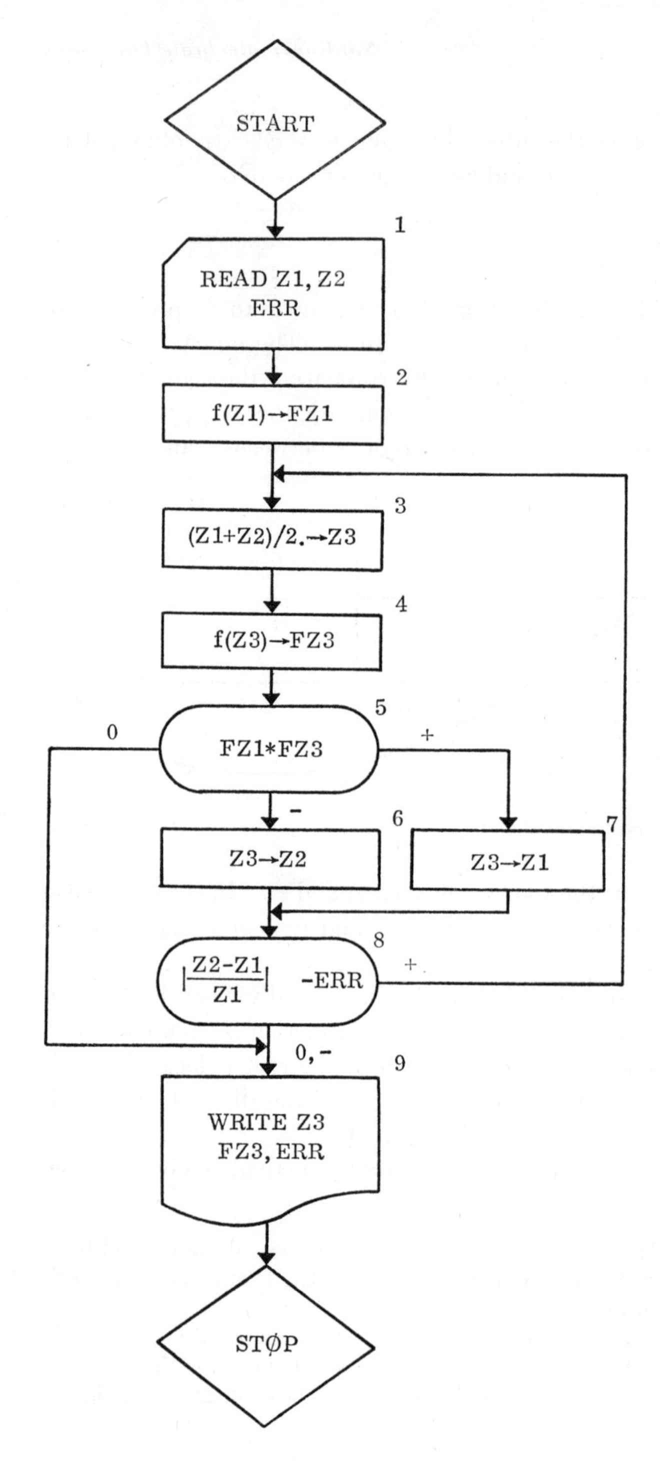

Figure 3.3 Flow chart for interval-halving method.

3,4. The value of the new approximation to the root is computed and stored in Z3, and the value of $f(z_3)$ is computed and stored in FZ3.

5–7. A comparison of the signs of FZ1 and FZ3 is made by testing the product of the two functions. If the signs are the same (product is positive), the root is between z_3 and z_2. z_3 is then placed in location Z1. If the signs are different (product is negative), the root is between z_1 and z_3. z_3 is then placed in location Z2. This will ensure that the root is always located between the numbers currently in Z1 and Z2.

8. A test is made on the ratio of the length of the interval $z_2 - z_1$ compared with z_1. If the ratio is greater than the admissible value in ERR, control is returned for another iteration. If it is less, the results are written out.

The rate of convergence of the interval-halving method is determined directly from the way in which each new approximation is made. After k iterations the interval within which the root is located will be $(\frac{1}{2})^k$ times the original interval. The kth approximation can have an error of no more than $(\frac{1}{2})^k$ times the initial interval on z. Conditions which the function $f(z)$ must satisfy for the success of this technique are that it be single-valued and that one and only one real root exist in the initial interval.

Because of the ease of programming this method and its generality for finding real roots, it is a useful technique. It is readily expanded so that more than one root of the function can be found by adding a stepwise search that will identify a change in sign of the function. When the function changes sign between two successive values for the argument, it has a zero in this interval, provided, of course, that it is continuous.

The flow chart in Fig. 3.4 shows the interval-halving routine combined with a stepwise search to find real roots of $f(z) = 0$ on the interval $z = a$ to $z = b$. Statements 1 to 8 accomplish the search for the interval within which the curve crosses the axis. Statements 9 to 17 perform the interval-halving routine in a slightly different form from the flow chart in Fig. 3.3. An explanation is given for statements 1 to 8 in the stepwise search.

1. The interval of the search a and b, the step size for the search, and the admissible error in the root are read into locations A, B, DELZ, and ERR, respectively.

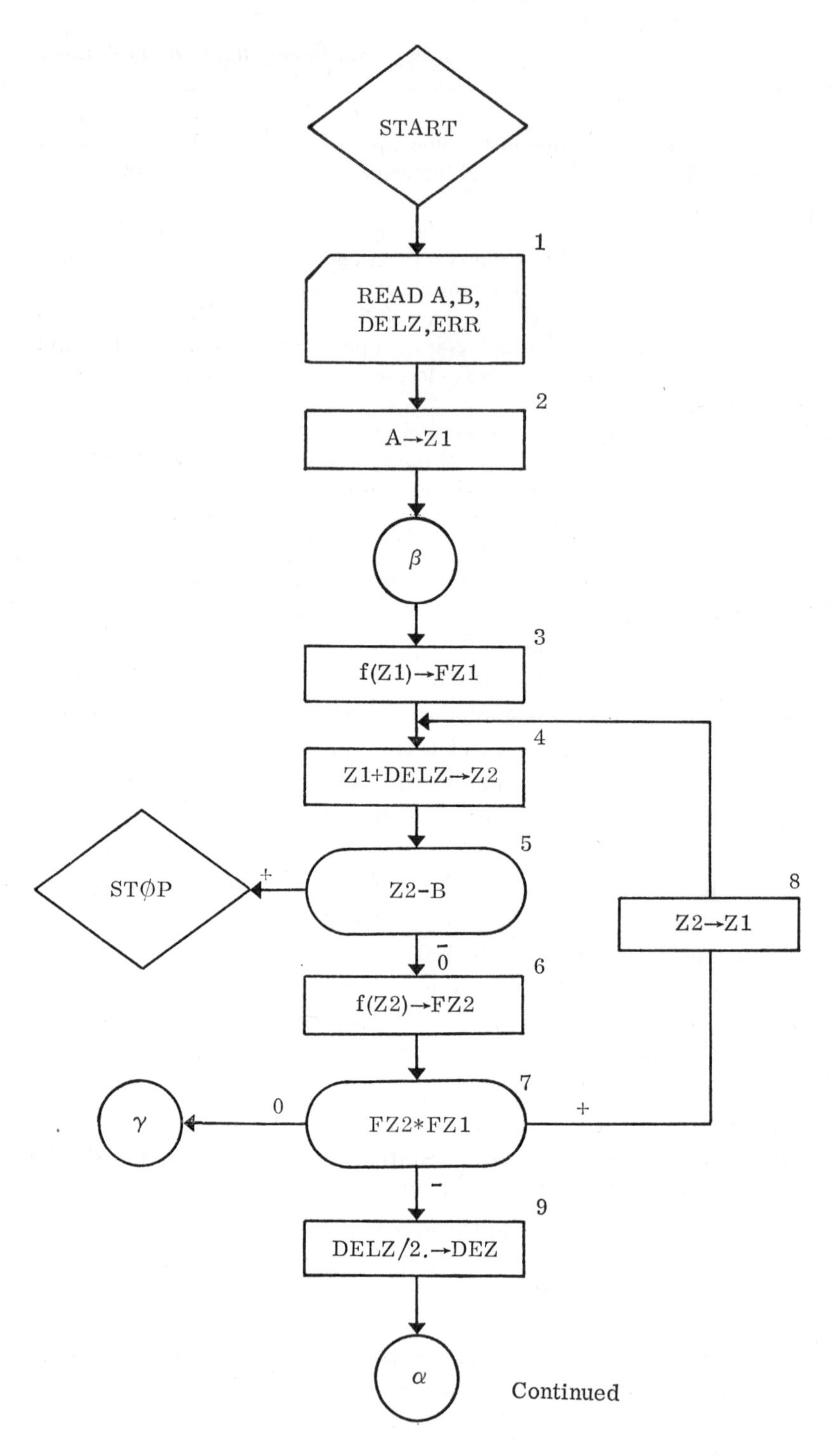

Figure 3.4 Flow chart to find several roots by interval-halving method.

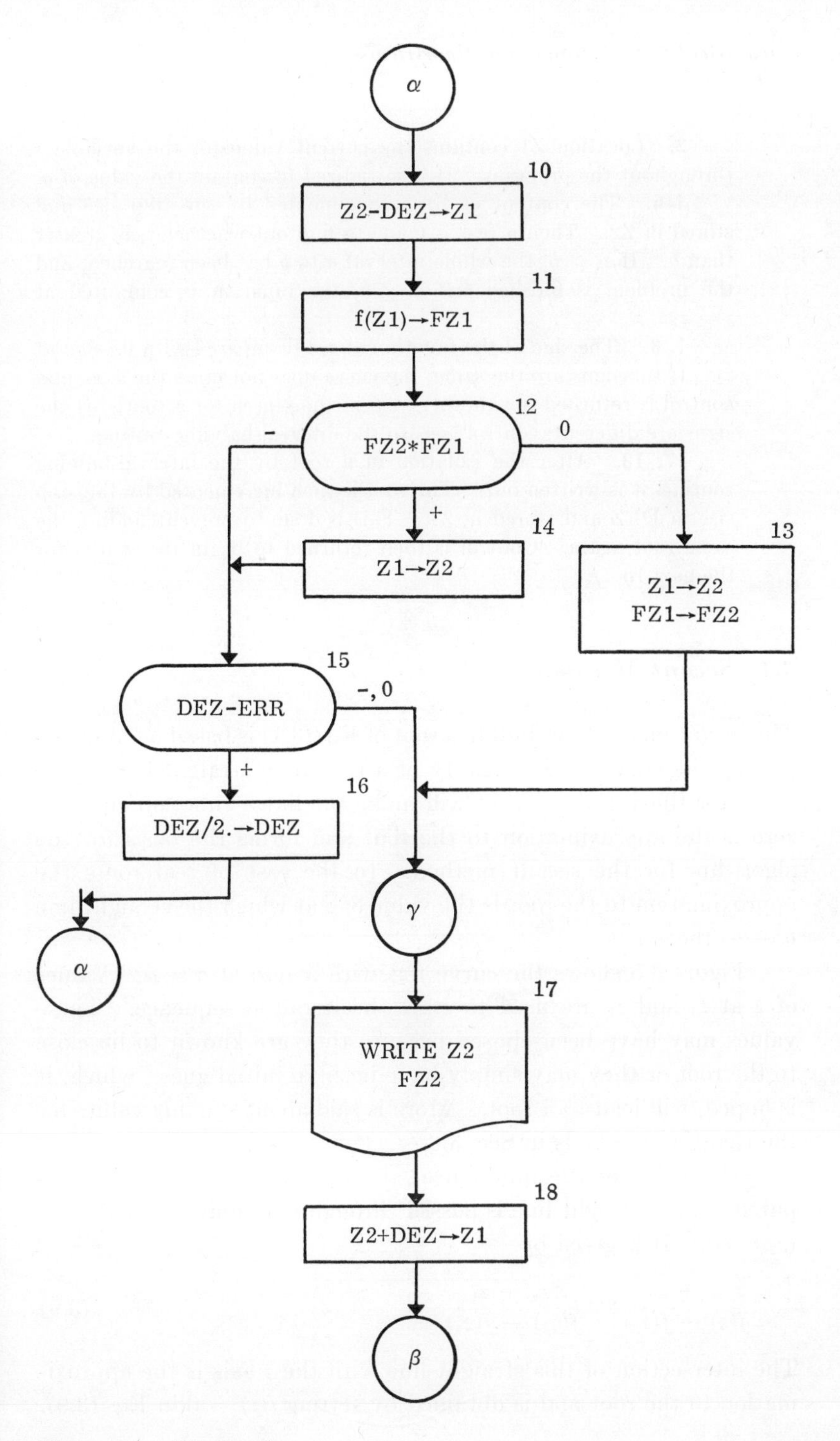

Figure 3.4 *(Continued)*

2. Location Z1 contains the current value for the variable z throughout the program. It is initialized to contain the value of a.

4–6. The content of Z1 is incremented by the step size and stored in Z2. Then a test is made to find out whether z_2 is greater than b. If $z_2 > b$, the whole interval a to b has been searched, and the problem is finished. If $z_2 \leq b$, the function is computed at $z = z_2$.

7, 8. The sign of the function at z_1 is compared with its sign at z_2. If the signs are the same, the curve does not cross the axis, and control is returned for another step in the search for a root. If the signs are different, control goes to the interval-halving routine.

17, 18. After the isolation of a root by the interval-halving routine, it is written out. The root is then incremented by the step size in DEZ and stored in Z1. This is done to prevent finding the same root again. Control is then returned to begin the search for the next root.

3.4 Secant Method

The secant method for finding a root of Eq. (3.1) is based on approximating the curve in the vicinity of a root by a straight line. The value for the variable z that will make the linear function equal to zero is the approximation to the root and forms the basis for the algorithm for the secant method. In the case of real roots the approximation to the root is the value of z at which the straight line crosses the z axis.

Figure 3.5 shows the curve $f(z)$ with a zero at $z = \alpha$. Values of z at z_1 and z_2 are used to start the iteration sequence. These values may have been chosen because they are known to lie close to the root or they may simply have been an initial guess, which, it is hoped, will lead to a root. More is said about starting values for the iteration methods in Sec. 3.6.

The value of the function at the two starting values is computed, and a straight line is passed through the points $(z_1,f(z_1))$ and $(z_2,f(z_2))$. It is given by

$$\frac{z - z_2}{f(z) - f(z_2)} = \frac{z_2 - z_1}{f(z_2) - f(z_1)} \tag{3.9}$$

The intersection of this straight line with the z axis is the approximation to the root and is obtained by setting $f(z) = 0$ in Eq. (3.9).

This gives the basic iteration formula

$$z = z_2 - \frac{(z_2 - z_1)f(z_2)}{f(z_2) - f(z_1)} \tag{3.10}$$

for the secant method.

The approximation obtained from Eq. (3.10) is the value z_3 in Fig. 3.5. It is used with the point at z_2 to find a new straight-line approximation to the curve that passes through the points $(z_2,f(z_2))$ and $(z_3,f(z_3))$. The intersection of the line with the z axis gives the

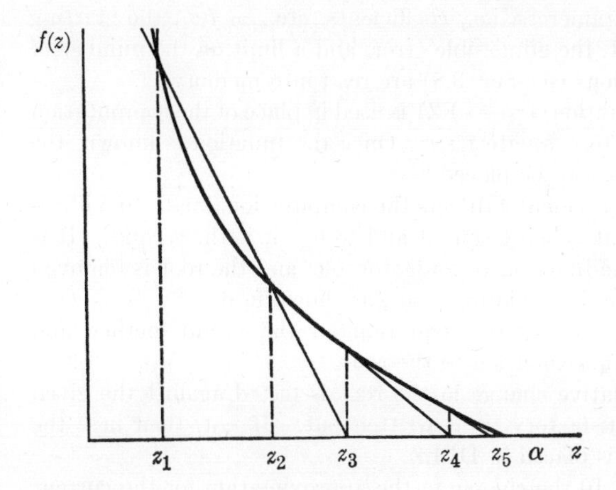

Figure 3.5 The secant method.

next approximation to the root designated z_4 in Fig. 3.5. The procedure of using the two most recently determined points in order to find a new iterate is continued until a satisfactory value is obtained or until it is established that the process is not convergent.

The secant method may not converge to a root in the vicinity of the initial guess. Such a possibility may occur for the curve in Fig. 3.8. When this happens, it is sometimes sufficient to alter the initial values z_1 and z_2 used to start the iteration. If this does not prove fruitful, one of the other methods in this chapter should be tried.

The ease of programming the secant method, its ready adaptation for finding complex roots as well as real roots, and its generally good convergence properties make it competitive with the other methods presented here. It is especially useful for nonpolynomial forms where the Newton-Raphson and Lin-Bairstow methods cannot be used. The secant method has many of the same advantages as the Müller method (Sec. 3.5), but in general it will not converge so rapidly.

A flow chart for finding two roots of Eq. (3.1) is given in Fig. 3.6. It is assumed that the two roots are real and that the function cannot be factored. An explanation of pertinent steps in the flow chart is given.

1. The parameters, i.e., coefficients, etc., in $f(z)$, the starting values for a root, the admissible error, and a limit on the number of diverging iterations (see Sec. 3.8) are read into memory.

3. The notation $f(z_1) \rightarrow$ FZ1 is used in place of the computation steps necessary to evaluate $f(z_1)$. Once the function is known, the appropriate steps can be placed here.

4. Logic statement 4 directs the computational path to 5 if it is the first root that is being sought and to 6 if it is the second. It is assumed that the function is nonfactorable, and the root is removed from the function by dividing it out, as shown in 6.

7. This is the iteration equation for the secant method and gives the new approximation to the root.

8. The relative change in the root is tested against the given error, and if satisfactory, it is written out. If not, then in 9 the relative change is placed in DELZ.

10, 11. In 10 the change in the approximation for the current computation is compared with the change for the previous computation. If the change is greater, it is counted in L. In either event the content of DEZ is updated to equal the change in the last computation.

12. The number of times that the change represented in 9 increases, i.e., L, is compared with a given limiting value contained in location LIMIT. This is intended to identify a diverging sequence or a condition where ERR has been made too small for the problem. When L equals the limiting value, the iterations are stopped.

14. The contents of Z1, Z2, and FZ1 are updated for another iteration, and control is sent to statement 4.

15. If the root is acceptable on the basis of the test in 8, it is written out.

16, 17. The number in N is tested against 1 to find out whether two roots have been found. If only one root is obtained, N contains zero, and control is sent to 17 to update N and finally to statement 4 for the second root.

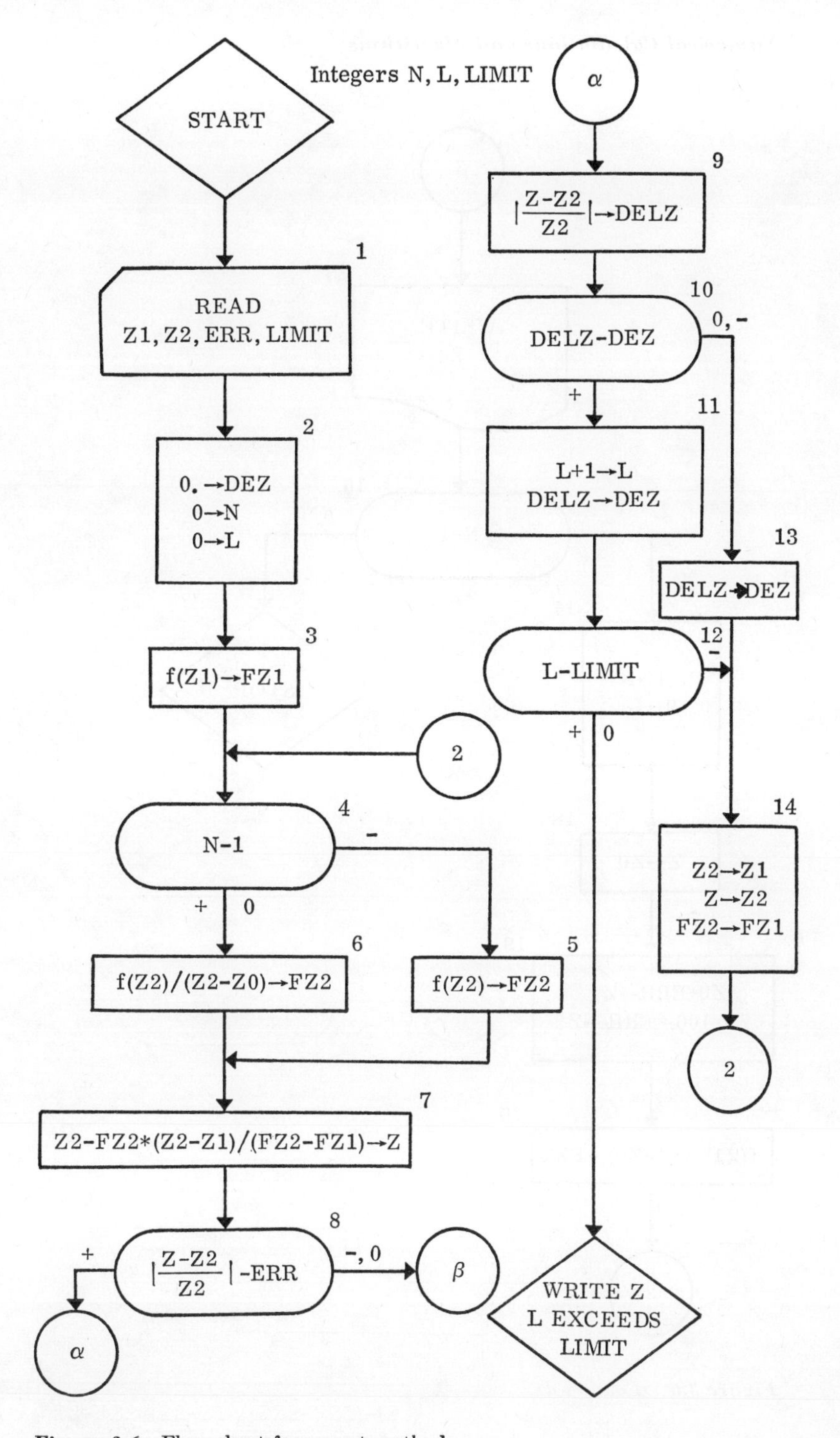

Figure 3.6 Flow chart for secant method.

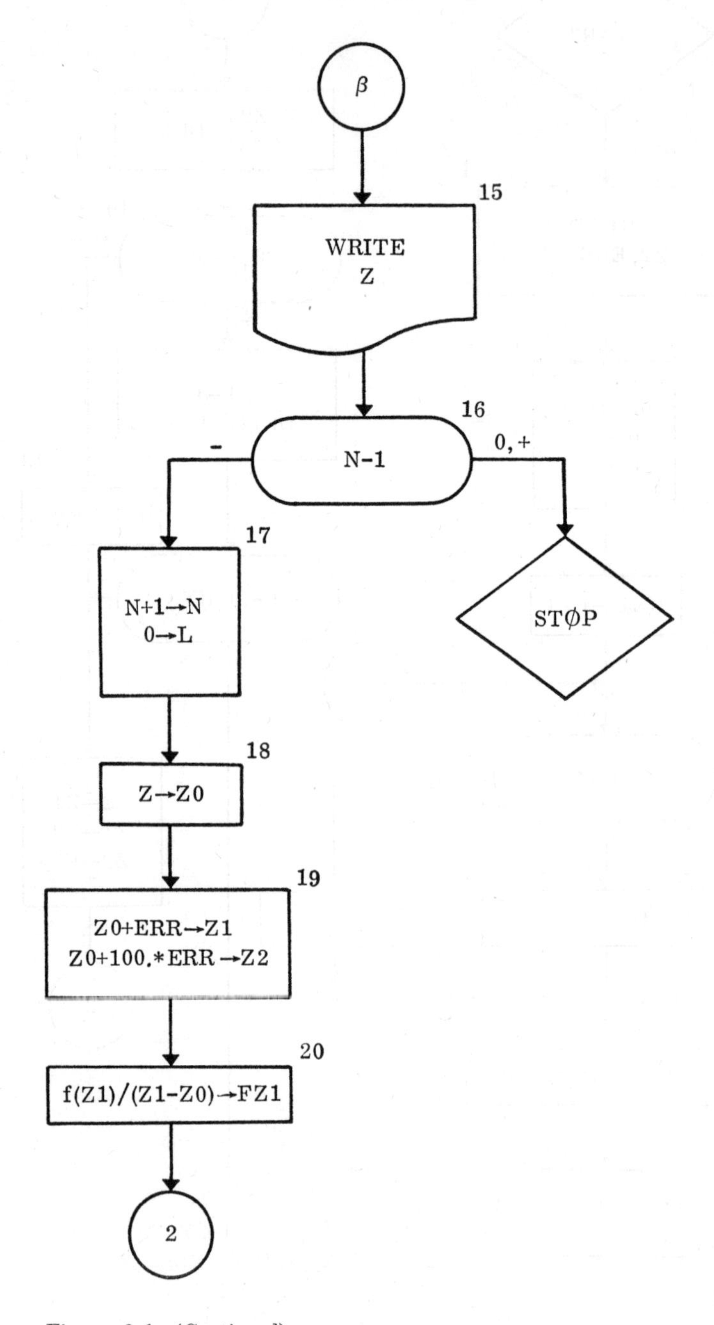

Figure 3.6 *(Continued)*

18–20. The starting values for the second root are obtained by adding small increments to the root already obtained. In 20 the value of the function at z_1 divided by the factor $(z_1 - z_0)$ is computed and placed in FZ1. If the function is a polynomial, it is factored at this point to obtain the reduced polynomial. (See the flow charts for the Newton-Raphson and the Lin-Bairstow methods.)

Example 3.1

Find two roots of the following polynomial by using the method outlined in Fig. 3.6:

$$f(z) = z^4 - 2z^3 + z^2 + 2z - 2$$

The initial approximations $z_1 = 0.0$ and $z_2 = 0.01$ were used to give the computer results in Table 3.1. A value of 10^{-5} was used for ERR.

Table 3.1

i	First root		Second root	
	z	$f(z)$	z	$f(z)/z - z_0$
1	0.00	−2.0	1.00010	2.000000
2	0.01	−1.979902	1.01000	2.010198
3	0.995123	−0.009730	−0.941390	0.279509
4	0.999988	−0.000023	−1.256543	−1.562858
5	1.000000	−0.000000	−0.989203	0.053521
6			−0.998055	0.009710
7			−1.000017	−0.000085
8			−1.000000	0.000000

Three iterations were sufficient for the first root at $z = z_0 = 1.0$. Starting values for the second root were obtained from

$$z_1 = z_0 + 10 \text{ ERR}$$
$$z_2 = z_0 + 1{,}000 \text{ ERR}$$

Six iterations were required to obtain the second root to a satisfactory approximation.

3.5 Müller Method

The Müller method is similar in concept to the secant method in that the curve $f(z)$ is replaced by a simpler form. In this method the approximation is a quadratic equation, which generally gives faster convergence than the secant method and thus gives an acceptable approximation to a root in fewer iterations. The computer time to complete an iteration by the Müller method is not significantly greater than for the other methods presented in this chapter. For this reason, and because of its excellent convergence properties, it is among the most successful methods for finding roots on an automatic computer. The presentation of the method follows closely Müller's original paper.[5]*

Let z_1, z_2, and z_3 be three points that are expected to be in the vicinity of a root. If an approximation to the root is available, let it be z_3; then

$$\begin{aligned} z_2 &= z_3 + \delta \\ z_1 &= z_3 - \delta \end{aligned} \tag{3.11}$$

where δ is an estimate of the maximum error in z_3. If no information about a root is known, the initial values $z_3 = 0$, $z_2 = 1$, $z_1 = -1$ are suggested.

The value of $f(z)$ at each of the starting values z_1, z_2, and z_3 is computed, and the quadratic which passes through these three points is constructed. It is easily verified that $y(z)$ in Eq. (3.12) is the required quadratic.

$$\begin{aligned} y(z) = \frac{(z - z_2)(z - z_3)}{(z_1 - z_2)(z_1 - z_3)} f(z_1) &+ \frac{(z - z_1)(z - z_3)}{(z_2 - z_1)(z_2 - z_3)} f(z_2) \\ &+ \frac{(z - z_1)(z - z_2)}{(z_3 - z_1)(z_3 - z_2)} f(z_3) \end{aligned} \tag{3.12}$$

Equation (3.12) is an approximation to $f(z)$ in the vicinity of $z = z_3$. It is assumed that the roots of $y(z)$ will be approximations to the roots of $f(z)$ in the region of the approximation. In particular the root of $y(z) = 0$ that is nearest to z_3 is chosen as the new approximation to the root of $f(z) = 0$. This root is designated z_4 and is then used in place of z_1 in a new quadratic approximation to $f(z)$. The new quadratic is in turn solved for the root closest to z_4, and the procedure repeated all over again. This process is continued until a satisfactory approximation is obtained.

* Superscript numbers pertain to references at the end of the chapter.

A change of variables makes the solution of Eq. (3.12) for the root nearest to z_3 a much simpler computational problem and also gives much improved convergence properties. Let

$$\lambda = \frac{z - z_3}{z_3 - z_2} \qquad \lambda_1 = \frac{z_3 - z_2}{z_2 - z_1} \qquad \delta_1 = \frac{z_3 - z_1}{z_2 - z_1} \tag{3.13}$$

Then Eq. (3.12) transforms into the following quadratic in λ:

$$\lambda^2\lambda_1\delta_1^{-1}[\lambda_1 f(z_1) - \delta_1 f(z_2) + f(z_3)] + \lambda\delta_1^{-1}[\lambda_1^2 f(z_1) \\ - \delta_1^2 f(z_2) + (\lambda_1 + \delta_1)f(z_3)] + f(z_3) = y(\lambda) \tag{3.14}$$

Equation (3.14) is solved for its roots by solving for $1/\lambda$ and then inverting the result.

$$\lambda = \frac{-2\delta_1 f(z_3)}{g_1 \pm \sqrt{g_1^2 - 4\delta_1 c_1 f(z_3)}} \tag{3.15}$$

where

$$\begin{aligned} g_1 &= \lambda_1^2 f(z_1) - \delta_1^2 f(z_2) + (\lambda_1 + \delta_1) f(z_3) \\ c_1 &= \lambda_1[\lambda_1 f(z_1) - \delta_1 f(z_2) + f(z_3)] \end{aligned} \tag{3.16}$$

The sign in the denominator of Eq. (3.15) is chosen to make λ have the smallest absolute value. The new approximation to the root is

$$z = z_3 + \lambda(z_3 - z_2) \tag{3.17}$$

If z is a satisfactory approximation to the root, it is recorded. If not, for the new iteration z_2 will replace z_1, z_3 will replace z_2, and the new approximation to the root in Eq. (3.17) will replace z_3. Another iteration is then made using Eqs. (3.15) and (3.17).

The computational steps in the Müller method are outlined below.

1. Select starting values for z_1, z_2, and z_3.

2. Compute $f(z_1)$, $f(z_2)$, and $f(z_3)$ and the values for λ_1 and δ_1 from Eqs. (3.13).

3. Compute values for g_1 and c_1 according to Eqs. (3.16) and then find the value for λ from Eq. (3.15). The sign in the denominator is chosen to agree with the sign of g_1.

4. The new approximation to the root is obtained from Eq. (3.17).

5. Test the approximation to find whether it meets the required error criterion. The recommended test is on the absolute value of the ratio of the change in the approximation to the approximation itself. Thus

$$\left| \frac{z - z_3}{z_3} \right| = \left| \lambda \frac{z_2 - z_3}{z_3} \right| < \epsilon$$

where ϵ is the given error criterion.

It is also suggested that the test be made in such a way as to detect a lack of improvement in the iteration. Thus if the ratio above is not decreasing with each iteration, the limit of convergence may already have been reached, and further computations would not improve the approximation. In such an event the iterations should be stopped and this fact noted in the print-out of results.

If the error criteria are not satisfied and further improvement in the root is possible, the iteration should be repeated for another approximation.

6. Before the next iteration is made, the values of z_3, z_2, and z_1 must be changed (updated) as follows:

$$z_2 \longrightarrow z_1$$
$$z_3 \longrightarrow z_2$$
$$z \longrightarrow z_3$$

Also the values of $f(z)$ must be obtained for the new set z_1, z_2, and z_3. Since $f(z_1)$ and $f(z_2)$ are available from the previous iteration, only $f(z_3)$ has to be calculated.

7. Steps 3 to 6 are repeated until the test in 5 indicates that the iterations should not be repeated again, at which time the results are written out.

Two numerical examples are given to illustrate this method.

Example 3.2

Find a root of $f(z)$:

$$f(z) = z \sin z + \cos z$$

The initial choices for z_3, z_2, and z_1 are obtained by using Eqs. (3.11) with $z_3 = 2$ and $\delta = 1$. The results of the iterations are shown in Table 3.2.

Table 3.2
A Root of $f(z) = z \sin z + \cos z$ by the Müller Method

k	z_k	$f(z_k)$	λ_k	δ_1	g_1	c_1
1	1.0000000	1.3817733				
2	3.0000000	−0.5666325	−0.5000000			
3	2.0000000	1.4024480	−0.7946749	0.5000000	0.4871014	−0.4974388
4	2.7946749	0.0097666	0.0047367	0.2053251	−0.4227139	−0.1367615
5	2.7984390	−0.0001395	−0.0140639	1.0047366	−0.0099687	−0.0000157
6	2.7983860	0.0000000	−0.0001068	0.9859305	0.0001376	−0.0000000
7	2.7983860					

Example 3.3

Find a root of $f(z)$:

$$f(z) = z^4 - 2z^3 + z^2 + 2z - 2$$

The initial choices for z_3, z_2, and z_1 are 0,0.01, and −0.01, respectively. The results of the interations are shown in Table 3.3. The entries in

Table 3.3
A Root of $(f)z = z^4 - 2z^3 + z^2 + 2z - 2$ by the Müller Method

k	z_k	$f(z_k)$	λ_k	δ_1	g_1	c_1
1	−0.010000	−2.019898				
2	0.010000	−1.979902	−0.50000			
3	0.000000	−2.000000	−73.20445	0.50000	−0.01000	0.000050
4	0.732045	−0.497434	0.32206	−72.20445	−110.78365	−2.291452
5	0.967809	−0.063411	0.15097	1.32206	0.55774	−0.016070
6	1.003402	0.006815	−0.09628	1.15097	0.08154	0.000710
7	0.999975	−0.000050	−0.00734	0.90372	−0.00619	0.000010
8	1.000000	−0.000000	0.00030	0.99266	0.00005	0.000000
9	1.000000					

Tables 3.2 and 3.3 are self-explanatory except for λ_k, which gives the values for both λ and λ_1 in the computations. The first entry for λ_k opposite $k = 2$ is λ_1 in the first iteration. Opposite $k = 3$ is the value for λ from the first iteration, which is λ_1 in the second iteration, etc.

These examples illustrate the relatively fast convergence of the Müller method. In Example 3.2 the method is applied to a transcendental equation where the initial approximation to a root is fairly close, whereas in the second example the starting values are removed from the root. It is interesting to note that the secant method of the previous section gives faster convergence to the root of the polynomial in Example 3.3. This is not completely unexpected, since convergence in any given case is dependent upon starting values as well as on the method. The starting values for z_1 and z_2 in Table 3.1 are especially well chosen for the root at $z = 1$. In general, the Müller method has better convergence than the secant method.

3.6 Newton-Raphson Method

The Newton-Raphson method is widely accepted as one of the best methods for solving for the roots of Eq. (3.1). The excellent results that are generally obtained with the method and the simple computational routine justify its popularity. The method applies as well for complex roots as for real roots, and the iterations converge rapidly provided the initial estimate for a root is close enough.

The algorithm for the Newton-Raphson method is obtained from a Taylor-series expansion of $f(z)$ about an approximation to a root. Let $z = z_0$ be an estimate to a root α. Then

$$f(z) = f(z_0 + h) = f(z_0) + hf'(z_0) + \frac{h^2}{2!}f''(\xi) \tag{3.18}$$

where ξ is on the range z_0 to $z_0 + h$.

If $z_0 + h$ is set equal to α, then

$$f(\alpha) = 0 = f(z_0) + hf'(z_0) + \frac{h^2}{2!}f''(\xi) \tag{3.19}$$

An estimate to the value of h can be made by using only the first two terms in Eq. (3.19). Let this estimate be designated h_1.

$$h_1 = -\frac{f(z_0)}{f'(z_0)} \tag{3.20}$$

The basic formula for the iterations in the Newton-Raphson method is obtained by adding h_1 to the estimate z_0. This new approximation is designated z_1.

$$z_1 = z_0 + h_1 = z_0 - \frac{f(z_0)}{f'(z_0)} \tag{3.21}$$

The $(k + 1)$st approximation to the root is obtained by using the kth approximation in the right-hand side of Eq. (3.22):

$$z_{k+1} = z_k - \frac{f(z_k)}{f'(z_k)} \tag{3.22}$$

The iteration defined by Eq. (3.22) usually gives fast convergence to a root of $f(z) = 0$ provided the error in the initial approximation z_0 is small. Good results can even be obtained when the initial approximation is not close to a root, provided the slope on the interval between $z = z_0$ and $z = \alpha$ is not small. These statements are verified by the expression for the error in the first iterate*

$$E(z_1) = \alpha - z_1 = \frac{f''(\xi)}{2f'(z_0)} (\alpha - z_0)^2 + O(\alpha - z_0)^3 \tag{3.23}$$

Equation (3.23) says the error in the first approximation from Eq. (3.22) $(k = 0)$ is $O(h^2)$, where $h = \alpha - z_0$. For this reason the method is said to be quadratically convergent and is a second-order method. Convergence of the method depends critically on the size of $\alpha - z_0$. It also depends upon the magnitude of the coefficient $f''(\xi)/2f'(z_0)$. If this term is large, convergence is slow if it occurs at all. If it is small, the method may give a good approximation to the root even when the error in z_0 is not small. The dependence of the convergence of the method on the initial approximation and on the slope of the curve in the interval z_0 to α is illustrated in Figs. 3.7 and 3.8.

Figure 3.7 shows the geometric interpretation for the Newton-Raphson method when the root at α is real. Comparison with Fig. 3.5 shows the similarity between this and the secant method. In fact, the two figures illustrate a case where the secant method can give faster convergence. A value of $z = z_0$ is the initial estimate to the root at α. Through the point at $(z_0,f(z_0))$ a straight line is constructed which has a slope equal to the slope of the curve $f(z)$ at $z = z_0$. The intersection of this line with the z axis gives the next approximation to the root at $z = z_1$. This operation is equivalent to Eq. (3.21). At $z = z_1$ a new line is constructed through the point $(z_1,f(z_1))$ with a slope equal to $f'(z_1)$. This line gives the approximation z_2. The process is continued until a suitable

* See Ref. 9 for a derivation of this equation.

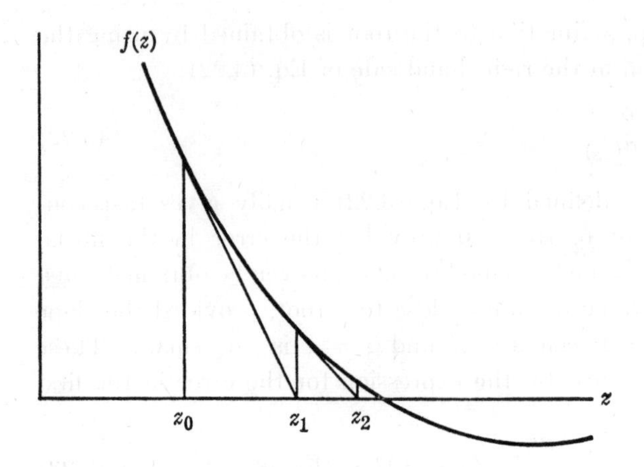

Figure 3.7 Newton-Raphson method.

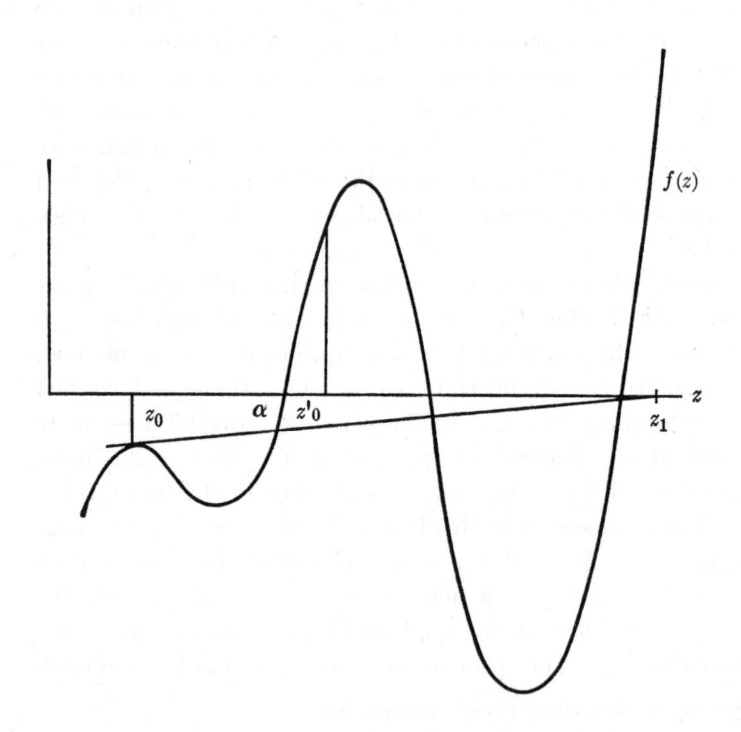

Figure 3.8 Starting values for Newton-Raphson method.

approximation is found. For the case in Fig. 3.7 convergence of the method is very fast, even when z_1 is not close to the root.

Figure 3.8 illustrates a potential difficulty with the Newton-Raphson method. The initial approximation at z_0 gives a very large value for the correction term h_1 because of the flat slope of the curve at $z = z_0$. In fact, the next value is moved far away from z_0, so that the process does not converge to the root at α. This is similar to the behavior of the secant method noted earlier. Clearly a better initial approximation, such as z_0', could give convergence to the root at $z = \alpha$.

Figures 3.7 and 3.8 point up the importance of the initial approximation in the Newton-Raphson method. The suggestions offered below for selecting starting values for this method can also be used for some of the other iterative schemes.

1. If $f(z)$ is a polynomial, an estimate of the smallest root can be made by truncating the higher-degree terms so that only a quadratic remains. The smallest root of the quadratic is an estimate of the smallest root of $f(z)$. An estimate of the largest root of a polynomial can be made in a similar way by truncating the lower-degree terms.

2. An estimate of the real roots can be made by the stepwise search, as in the flow chart in Fig. 3.4. This is especially effective for transcendental equations.

3. A small complex number can be used as a starting value for the iterations.

Suggestion 1 may lead to a complex number for an initial estimate, and suggestion 3 assumes a starting value that is complex. If either of these suggestions is used, the computation routine must be written for complex arithmetic.

If the starting value in the iteration routine is complex, the final value in the iteration sequence will be a complex number, regardless of whether the root is actually real or complex. This is because the iteration routine generally will not arrive at the exact value of the root. The identification of a real root is made on the basis of a comparison of the magnitude of the imaginary part of the root with the real part. If the imaginary part is significant when compared with the real part, the root is assumed to be complex. Otherwise it is assumed to be real. The measure of significance must be based upon the number of significant digits required for the root. For six-digit accuracy a ratio of the imaginary part to the real part of less than 10^{-3} is generally assumed to indicate a real root.

Example 3.4

The Newton-Raphson method is illustrated with the following transcendental equation:

$$f(z) = z \sin z + \cos z$$

The iteration equation for a root of this equation is obtained from (3.22).

$$z_{k+1} = z_k - \tan z_k - \frac{1}{z_k}$$

It is readily verified that the equation has a root between $z = \pi/2$ and $z = \pi$. Let $z_0 = \pi$ be a starting value for the iteration.

Successive iterations using five-digit numbers give the sequence 3.1416, 2.8233, 2.7986, 2.7984, 2.7984, ... The fourth value in the sequence is equal to the third and is the limiting value of the iteration. Note that a starting value of $z_0 = \pi/2$ would give a value of infinity for z_1. Obviously $z_0 = \pi/2$ could not be used as a starting value, and values for z_0 near to $\pi/2$ would not give satisfactory results.

Figure 3.9 is a flow chart for the Newton-Raphson method applied to a polynomial of degree n. The coefficients in the polynomial are assumed to be complex. A small complex number is used to start the iterations. Additional roots are found by factoring out the roots as they are found and then solving the reduced polynomial by the same routine. The starting value for the reduced polynomial is based on the last root that has been found. If the last root is complex, it is used as a starting value for the next iteration; if it is real, the starting value is the product of the real root and $1 + j$, where $j = \sqrt{-1}$. An explanation of the flow chart follows.

1. The data are read into the machine. The coefficients in the polynomial are $a_i = b_i + jc_i$. The real and imaginary parts are read into the vector locations AR and AI, respectively, so that

$$a_i = \mathrm{AR(I)} + j\mathrm{AI(I)} \qquad \text{for } i = 0, 1, 2, \ldots, n$$

2. Certain initializing steps are taken. The vector locations BR and BI are used in evaluating $f(z)$ and $f'(z)$. The initial guess for the root is $z_0 = x + jy$. Its real and imaginary parts are stored in X and Y, respectively, and n is stored in K.

3–5. The number in K is compared with 1. If K is equal to 1, the following linear equation is solved, and all the roots are written out:

$$a_0 z + a_1 = 0$$

6–8. The numbers needed to evaluate $f(z)$ and $f'(z)$ are computed. The method used is the method described in Example 2.4. The vectors BR and BI are the real and imaginary parts of the coefficients of $g(z)$, where

$$f(z) = [z - (x + jy)]^2 g(z) + Sz + R$$

The equations for $b_i = \text{BR(I)} + j\text{BI(I)}$ are repeated here for handy reference.

$$b_{-1} = 0$$

$$b_0 = a_0$$

$$b_i = a_i - ub_{i-1} - vb_{i-2} \qquad \text{for } i = 1, 2, \ldots, n - 1$$

$$S = b_{n-1}$$

$$R = a_n - vb_{n-2}$$

where

$$u = -2(x + jy) \qquad \text{and} \qquad v = (x + jy)^2$$

9. The value of $f(z)$ at $z = x + jy$ is computed and stored in FR and FI. The formula is

$$f(z) = Sz + R = b_{n-1}z + a_n - vb_{n-2}$$

10. The change in z is computed according to Eq. (3.22). Here $f(z) = \text{FR} + j\text{FI}$ and $f'(z) = b_{n-1} = \text{BR(K} - 1) + j\text{BI(K} - 1)$.

11, 12. The next approximation to the root is computed and stored in X and Y. This approximation is then tested, and the iterations are continued if it is not satisfactory.

13, 14. The imaginary part of the root is set to zero if the root appears to be a real root.

15–17. The root is stored in the proper elements of the vectors RR and RI and is then factored from the polynomial. The equations used are derived in Example 2.4. The coefficients of the factored polynomial are stored back into the vectors AR and AI.

18–20. Preparatory to finding the next root, K is reduced by 1, and the last root is multiplied by $1 + j$ if it is a real root. Control is then returned to find the next root.

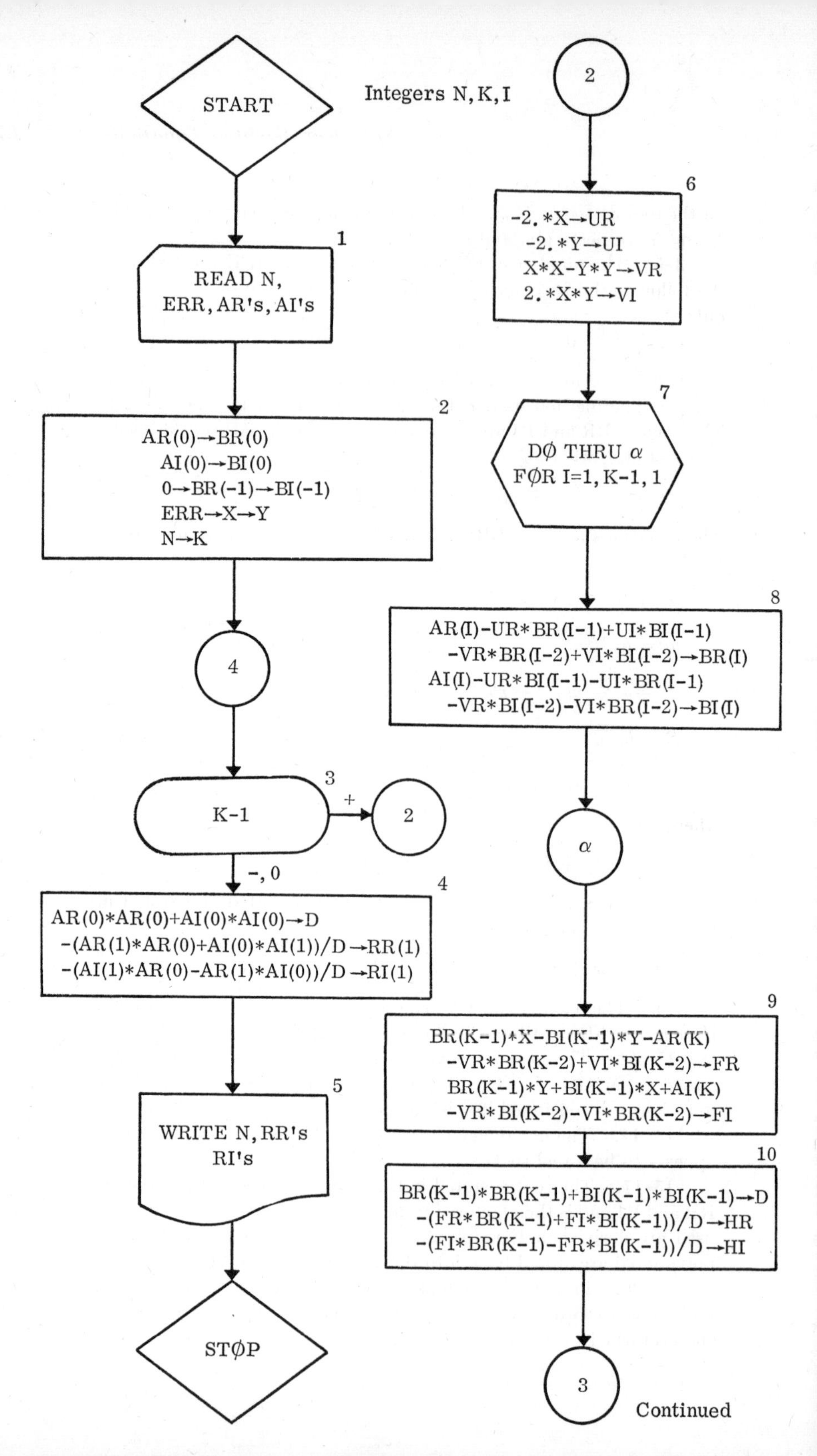

Figure 3.9 Roots of a polynomial by Newton-Raphson iteration.

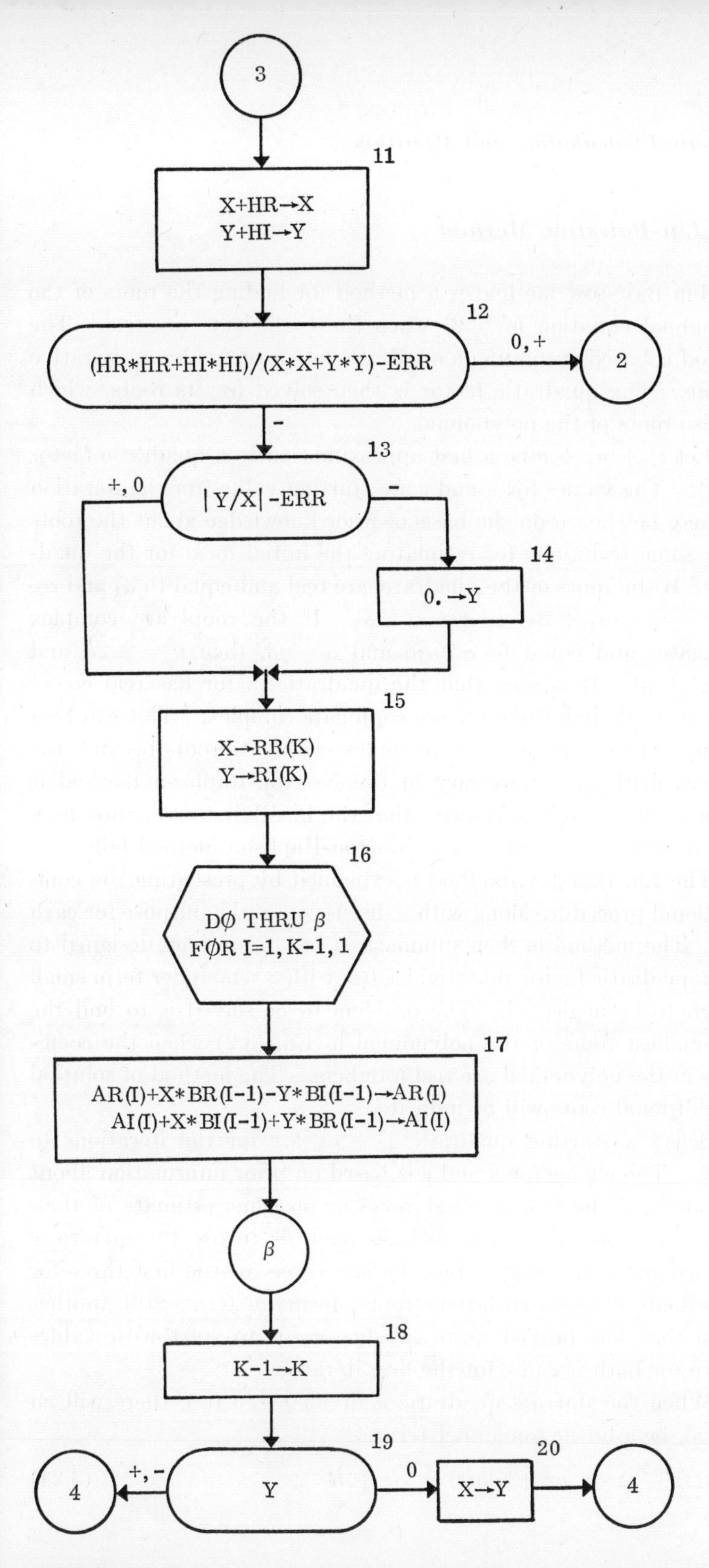

Figure 3.9 *(Continued)*

3.7 Lin-Bairstow Method

The Lin-Bairstow method is a method for finding the roots of the polynomial equation in (3.2) when the coefficients are real. The method is based on finding a quadratic factor of $f(z)$ by an iterative routine. The quadratic factor is then solved for its roots, which are also roots of the polynomial.

Let $z^2 + uz + v$ be a first approximation to a quadratic factor of $f(z)$. The values for u and v are starting values for the iteration and may be chosen on the basis of prior knowledge about the roots or by some technique for estimating the initial form for the quadratic. If the roots of the quadratic are real and equal to α_1 and α_2, then $u = -(\alpha_1 + \alpha_2)$ and $v = \alpha_1\alpha_2$. If the roots are complex conjugates and equal to $\alpha + j\beta$ and $\alpha - j\beta$, then $u = -2\alpha$ and $v = \alpha^2 + \beta^2$. It is seen that the quadratic factor has real coefficients even though the roots are conjugate complex. This can be a distinct aid in the solution of complex conjugate roots because the complex arithmetic necessary in the Newton-Raphson method is not needed. Also it is known[10] that the Lin-Bairstow method may yield results in cases where the Newton-Raphson method fails.

The Lin-Bairstow method is explained by presenting the computational procedure along with a discussion of the purpose for each step. The method is then summarized in a flow chart designed to find a quadratic factor that divides $f(z)$ with a remainder term small enough to be neglected. The problem to be solved is to find the two smallest roots of the polynomial in Eq. (3.2) when the coefficients in the polynomial are real numbers. The method of solution for additional roots will be indicated.

Select a starting quadratic $z^2 + uz + v$ for the iterations to follow. The choice for u and v is based on prior information about the values of the two smallest roots or on some estimate of their value. A choice that is sometimes made is to use the quadratic obtained by truncating all but the last three (or the first three for an estimate of the two largest roots) terms in $f(z)$. Still another choice that has proved quite satisfactory is to start with values of zero for both u and v for the first iteration.

When the starting quadratic is divided into $f(z)$, there will, in general, be a linear remainder term:

$$f(z) = (z^2 + uz + v)g(z) + Sz + R \tag{3.24}$$

The computational procedure for finding S and R and the $(n - 2)$nd-degree polynomial $g(z)$ is the synthetic division routine presented in Example 2.4.

$$
\begin{aligned}
b_0 &= a_0 \\
b_1 &= a_1 - ub_0 \\
b_i &= a_i - ub_{i-1} - vb_{i-2} \qquad i = 2, 3, \ldots, n-2 \\
S &= b_{n-1} = a_{n-1} - ub_{n-2} - vb_{n-3} \\
R &= a_n - vb_{n-2}
\end{aligned}
\tag{3.25}
$$

The object of the Lin-Bairstow method is to reduce the remainder terms S and R to as near to zero as required for a satisfactory approximation to the roots.

The remainder terms are dependent upon u and v; hence S and R are functions of u and v:

$$
\begin{aligned}
S &= S(u,v) \\
R &= R(u,v)
\end{aligned}
\tag{3.26}
$$

Consider a change in u and v represented by Δu and Δv. The new values for S and R can be given by the Taylor-series expansion about u and v. Only the first-order terms are given in the following equations:

$$
\begin{aligned}
S(u + \Delta u, v + \Delta v) &= S(u,v) + \frac{\partial S}{\partial u}\Delta u + \frac{\partial S}{\partial v}\Delta v \\
&\qquad + O(\Delta u^2 + \Delta v^2) \\
R(u + \Delta u, v + \Delta v) &= R(u,v) + \frac{\partial R}{\partial u}\Delta u + \frac{\partial R}{\partial v}\Delta v \\
&\qquad + O(\Delta u^2 + \Delta v^2)
\end{aligned}
\tag{3.27}
$$

Let Δu and Δv be chosen so that the left-hand side of Eqs. (3.27) vanishes. An approximation to the values for Δu and Δv is obtained by using only the linear terms:

$$
\begin{aligned}
\Delta u \frac{\partial S}{\partial u} + \Delta v \frac{\partial S}{\partial v} &= -S(u,v) \\
\Delta u \frac{\partial R}{\partial u} + \Delta v \frac{\partial R}{\partial v} &= -R(u,v)
\end{aligned}
\tag{3.28}
$$

The derivative coefficients and the right-hand side of Eqs. (3.28) must be evaluated in order to solve for Δu and Δv. The values of S and R for any given values of u and v are found from Eqs. (3.25). The derivative coefficients in Eqs. (3.28) are found by taking the derivative of Eq. (3.24) with respect to u and then with respect to v to give

$$0 = zg(z) + (z^2 + uz + v)\frac{\partial g}{\partial u} + z\frac{\partial S}{\partial u} + \frac{\partial R}{\partial u} \tag{3.29a}$$

$$0 = g(z) + (z^2 + uz + v)\frac{\partial g}{\partial v} + z\frac{\partial S}{\partial v} + \frac{\partial R}{\partial v} \tag{3.29b}$$

If these equations are rewritten by placing the known polynomials [see Eqs. (3.25)] $zg(z)$ and $g(z)$ on the left-hand side, the right-hand side of each equation represents the form of a synthetic division by the factor $z^2 + uz + v$ with the derivatives of S and R as the remainder terms.

$$\begin{aligned} zg(z) &= b_0z^{n-1} + b_1z^{n-2} + \cdots + b_{n-3}z^2 + b_{n-2}z \\ &= (z^2 + uz + v)(d_0z^{n-3} + d_1z^{n-4} + \cdots \\ &\qquad + d_{n-4}z + d_{n-3}) + zd_{n-2} + d_n \\ g(z) &= b_0z^{n-2} + \cdots + b_{n-3}z + b_{n-2} \\ &= (z^2 + uz + v)(d_0z^{n-4} + d_1z^{n-3} + \cdots \\ &\qquad + d_{n-5}z + d_{n-4}) + d_{n-3}z + d_{n-1} \end{aligned} \tag{3.30}$$

It is readily shown that the coefficients in the quotient polynomial are the same in each of Eqs. (3.30), and they are so noted. By comparing coefficients the following recursion formulas are obtained:

$$\begin{aligned} d_0 &= b_0 \\ d_1 &= b_1 - ud_0 \\ d_i &= b_i - ud_{i-1} - vd_{i-2} \qquad i = 2, 3, \ldots, n-2 \\ d_{n-1} &= b_{n-2} - vd_{n-4} \\ d_n &= -vd_{n-3} \end{aligned} \tag{3.31}$$

where

$$\frac{\partial S}{\partial v} = -d_{n-3} \qquad \frac{\partial S}{\partial u} = -d_{n-2}$$
$$\frac{\partial R}{\partial v} = -d_{n-1} \qquad \frac{\partial R}{\partial u} = -d_n \tag{3.32}$$

Substituting (3.32) into Eqs. (3.28) and solving for Δu and Δv gives

$$\Delta u = \frac{-d_{n-3}R + d_{n-1}S}{d_{n-2}d_{n-1} - d_{n-3}d_n}$$
$$\Delta v = \frac{-d_nS + d_{n-2}R}{d_{n-2}d_{n-1} - d_{n-3}d_n} \tag{3.33}$$

The correction terms Δu and Δv are then added to u and v, respectively, to give the improved values. A test on the size of Δu and Δv will tell whether the procedure should be repeated to find still better values for u and v. When suitable values for u and v are found, the quadratic $z^2 + uz + v$ is solved for its roots.

A flow chart for the Lin-Bairstow method is presented in Fig. 3.10. An explanation of the flow chart is given below.

1. The degree of the polynomial, the initial values for u and v, error criteria EU and EV, and the coefficients in $f(z)$ are read into the designated locations.

2. The value of a_0 is placed in locations B(0) and D(0), and zero is placed in locations B(−1) and D(−1).

3–5. The coefficients b_i in the polynomial $g(z)$ and the remainder term are computed using Eqs. (3.25). The value for S is stored in location B(N − 1) and the value for R in B(N).

6–8. The coefficients d_i, $i = 1, \ldots, n$, are computed using Eqs. (3.31). Values of d_i are stored in the vector D. Equations (3.32) show that the locations D(I), $i = n - 3, n - 2, n - 1$, and n, contain the numerical values for the derivatives $\partial S/\partial v$, $\partial S/\partial u$, $\partial R/\partial v$, and $\partial R/\partial u$, respectively.

9. Values for Δu and Δv are obtained using Eqs. (3.33) and are stored in locations DELU and DELV. These increments are then added to the current values of u and v to give new approximations for u and v.

10, 11. The increments Δu and Δv are tested against a prescribed error. The test is made on the basis of the percentage change in the variables u and v. A negative result on either test means that

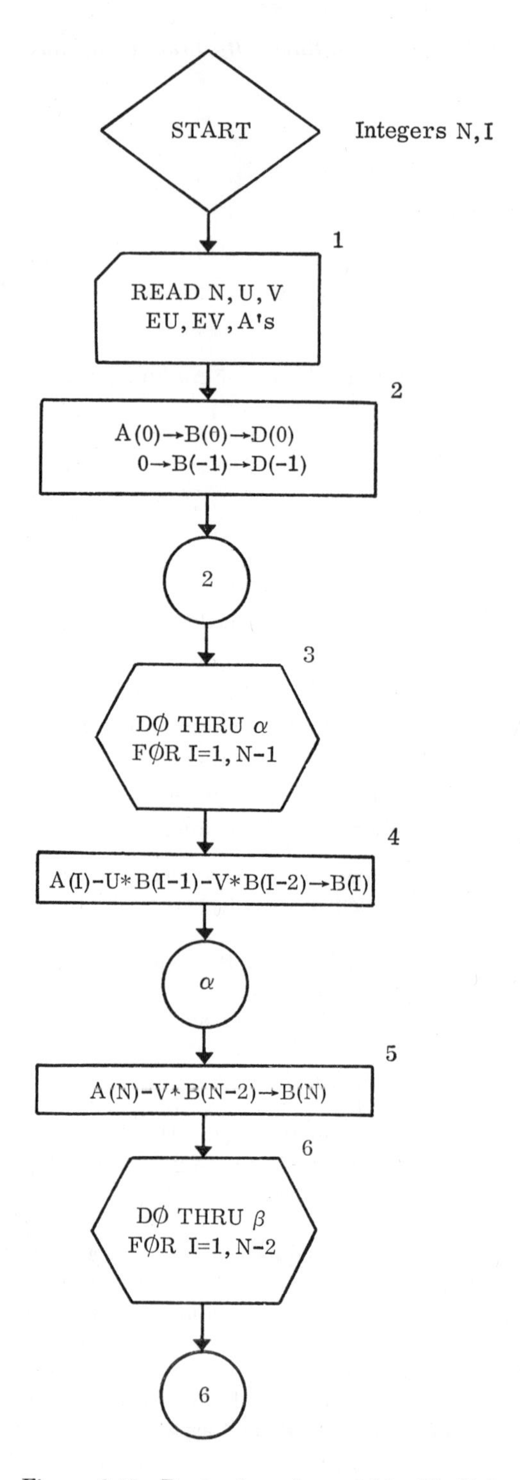

Figure 3.10 Roots of a polynomial by Lin-Bairstow method.

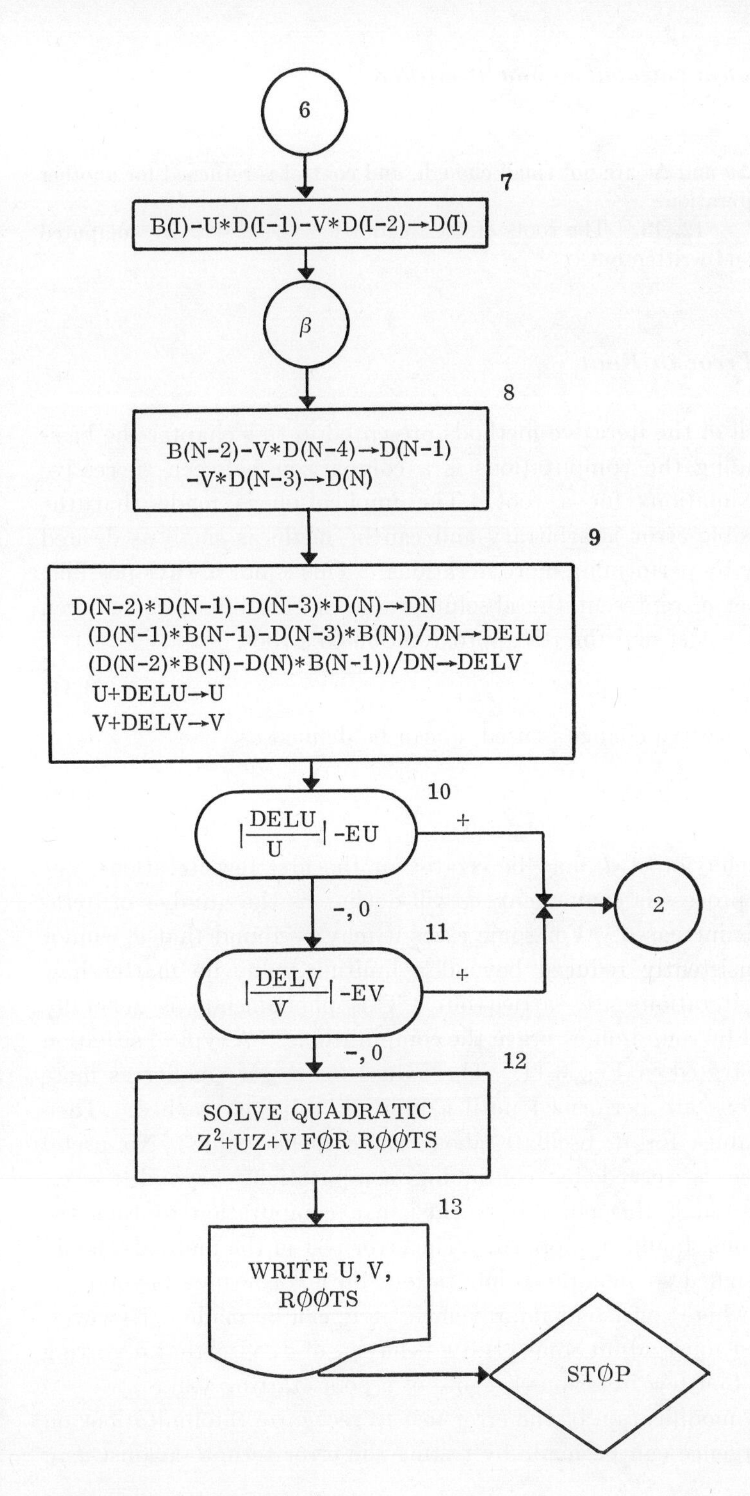

Figure 3.10 *(Continued)*

Δu and Δv are not small enough, and control is returned for another iteration.

12, 13. The roots of the quadratic $z^2 + uz + v$ are computed and written out.

3.8 Error in Root

In each of the iterative methods presented in this chapter the basis for ending the computations is a comparison between successive approximations for a root. The implication is made that the admissible error is arbitrary and can be made as small as desired simply by performing more iterations. This is not always possible.

Let d_i represent the absolute value of the difference between the $(i + 1)$st and the ith approximation to a root.

$$d_i = |z_i - z_{i-1}| \tag{3.34}$$

If the relative change is used, d_i can be defined as

$$d_i = \left| \frac{z_i - z_{i-1}}{z_i} \right| \tag{3.35}$$

The behavior of d_i may be erratic for the first few iterations, but if the process is convergent, d_i will decline as the number of iterations is increased. For some cases it may be found that d_i cannot be consistently reduced beyond a limiting value no matter how many iterations are carried out. This phenomenon is generally caused by round-off errors in the computations. A typical situation is illustrated in Fig. 3.11. The values for d_i get smaller as more iterations are performed until a limiting value is reached. Then the values for d_i oscillate about some mean value. No useful purpose is served by continuing computations beyond $i = n_0$. When and if this point is reached in a computation routine, the iterations should be stopped. The error test in the methods should be designed so that the computations do not continue beyond the point where consistent improvement in d_i can be made. However, the test must admit some erratic behavior of d_i, viz., that occurring in the first few iterations because of a poor starting value.

A modification of the error test to recognize the limitations on convergence can be made by testing the error term d_i against d_{i-1}.

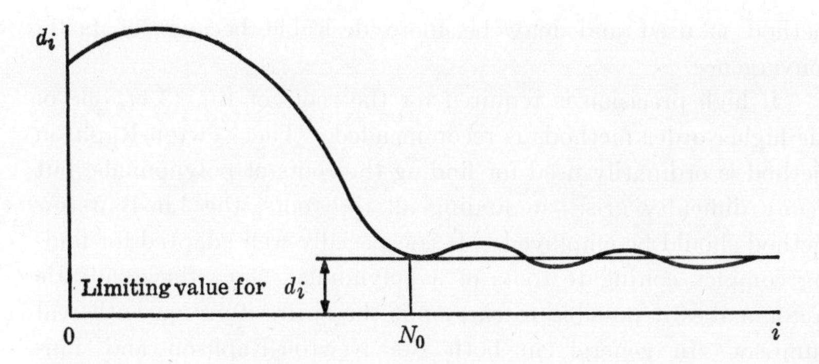

Figure 3.11 Typical behavior of d_i versus the number of iterations.

When there is no improvement over several iterations, the limit of accuracy has probably been reached. Performing more iterations will not give an improved approximation to the root. A procedure for such a test is outlined below.

1. Before beginning the iterations, place zeros into DOLD and L. DOLD contains the value of d_{i-1} during the ith iteration and L contains the number of times the value for d_i has increased from one iteration to the next.

2. After the new approximation to the root is computed, compute d_i and store it in DI.

3. Compare DI with DOLD. If DI $\geq$ DOLD, add 1 to L.

4. Place DI into DOLD and compare L with some small number, say 10. If L $<$ 10, return for another iteration. If L $\geq$ 10, the limiting value for d_i is assumed to be reached and the iterations should be stopped.

This test will also identify divergence of the iterations.

3.9 Comparison of Methods

The method of interval halving is a useful procedure for finding real roots of Eq. (3.1) when a high precision is not required. Satisfactory results have been obtained with the method for solving the characteristic equation for the frequencies of multi-degree-of-freedom systems. It has proved especially useful for transcendental equations where the roots are known to be real and positive. In general the secant method can be used where the interval-halving method is used and may be more desirable because of faster convergence.

If high precision is required for the roots of Eq. (3.1), one of the higher-order methods is recommended. The Newton-Raphson method is ordinarily used for finding the roots of polynomials, but if any difficulty arises in finding all the roots, the Lin-Bairstow method should be employed. It is especially well adapted for finding complex conjugate roots of a polynomial, since the quadratic factor makes it possible to carry out the entire iteration with real numbers. In general, in both the Newton-Raphson and Lin-Bairstow methods it is necessary to factor the roots from the function $f(z)$ as they are found, and for this reason these methods are not used for nonpolynomial forms where all the roots are required.

The Müller method is well adapted for application to nonpolynomial forms and is also recommended for polynomials with complex coefficients.

3.10 Summary

Methods for the numerical solution of nonlinear algebraic equations in one variable are presented. Only methods adapted for automatic computers are included, viz., the simple iteration, secant, interval-halving, Newton-Raphson, Lin-Bairstow, and Müller methods. Each method is based upon starting with some initial estimate of a root; then by an iterative procedure the estimate is improved. The methods of Newton-Raphson, Lin-Bairstow, and Müller, which are higher-order methods, in general give faster convergence to a root.

Techniques for estimating starting values for the iterations are suggested, and criteria for testing convergence are discussed.

Problems

1. Find a root of the following equations by the secant method. Obtain two significant figures. Use starting values of 0 and 1.

(a) $z^2 - 3z + 1 = 0$ (b) $z^3 - 5z + 3 = 0$

(c) $z^2 - \frac{1}{2} = 0$ (d) $-3z^3 - 2z + 4 = 0$

(e) $z^3 - \frac{1}{2} = 0$

2. (a) Do Prob. 1 using the method of interval halving. (b) Construct the flow chart.
3. (a) Do Prob. 1 using the Newton-Raphson method. Use a starting value of 0 where possible. (b) Construct the flow chart.
4. (a) Solve Prob. 1 using the Lin-Bairstow method with starting values of u, $v = 0$. (b) Construct the flow chart.
5. (a) Solve Prob. 1 using the Müller method with starting values of $(-1,0,1)$. (b) Construct the flow chart.
6. Compare the five methods used in Probs. 1 to 5 for the roots of Prob. 1b.
7. Construct a flow chart to find the square root of a ($a \geq 0$) by the Newton-Raphson method. *Note:* $z = \sqrt{a}$ is a root of the equation $z^2 - a = 0$.
8. Construct a flow chart to find the nth root of a ($a \geq 0$) by the Newton-Raphson method.
9. Use the Newton-Raphson method to find the smallest positive root of the following equations: (use $z_0 = 0$, $z_k > 0$)
   (a) $z - \cos z = 0$ (b) $z \sin z - \cos z = 0$
   (c) $z \tan z + 1 = 0$ (d) $\sinh z \cos z + 1 = 0$
10. Construct a flow chart to solve the set of equations in Prob. 9 for the smallest positive root by the secant method.
11. Construct a flow chart to solve the set of equations in Prob. 9 for the smallest positive root by the Müller method.
12. Construct a flow chart to solve Prob. 9b for the five smallest positive roots by the Newton-Raphson method. *Note:* Use a search routine to locate approximate value of a root.
13. Find a complex root of the following equations using the secant method. At least one of the initial guesses must be a complex number.
    (a) $z^2 - 2z + 2 = 0$ (b) $z^4 + 4 = 0$
    (c) $z^4 - z^3 + z^2 + 2 = 0$
    (d) $z^4 + 28z^3 + 70z^2 + 28z + 1 = 0$
    (e) $z^3 + 7z^2 + 7z + 1 = 0$
14. Write a flow chart to solve Prob. 13.
15. Find a complex root of the equations in Prob. 13 using the Newton-Raphson method.
16. Write a flow chart to solve Prob. 15.
17. Find a complex root of the equations in Prob. 13 using the Müller method.
18. Write a flow chart to solve Prob. 17.
19. Find the roots of the equations in Prob. 13 using the Lin-Bairstow method.
20. Write a flow chart to solve Prob. 19.

21. Write a flow chart to find a root of an nth-degree polynomial with complex coefficients by the secant method.

22. Do Prob. 21 using the Müller method.

23. Write a flow chart to find all of the roots of the equations in Prob. 13 by the secant method.

24. Write a flow chart to find all of the roots of the equations in Prob. 13 by the Müller method.

References

1. Henrici, Peter: "Elements of Numerical Analysis," John Wiley & Sons, Inc., New York, 1964.
2. Hildebrand, F. B.: "Introduction to Numerical Analysis," McGraw-Hill Book Company, New York, 1956.
3. Householder, A. S.: "Principles of Numerical Analysis," McGraw-Hill Book Company, New York, 1953.
4. Kunz, K. S.: "Numerical Analysis," McGraw-Hill Book Company, New York, 1957.
5. Müller, David: A Method for Solving Algebraic Equations Using an Automatic Computer, *Mathematical Tables and Computations*, vol. 10, October, 1956.
6. Olver, F. W. J.: Evaluation of Zeros of High Degree Polynomials, *Phil. Trans. Roy. Soc. London Ser. A*, vol. 244, pp. 385–415, 1952.
7. Ostrowski, A. M.: "Solutions of Equations and Systems of Equations," Academic Press Inc., New York, 1960.
8. Ralston, Anthony: "A First Course in Numerical Analysis," McGraw-Hill Book Company, 1965.
9. Scarborough, J. B.: "Numerical Mathematical Analysis," 5th ed., The Johns Hopkins Press, Baltimore, 1962.
10. Wilkinson, J. H.: The Evaluation of the Zeros of Ill Conditioned Polynomials, I, II, *Numer. Math.*, vol. 1, pp. 150–180, 1959.

Chapter 4

Simultaneous Linear Equations, Determinants, and Matrices

4.1 Introduction Equations of the form

$$
\begin{aligned}
a_{11}x_1 + a_{12}x_2 + \cdots + a_{1n}x_n &= b_1 \\
a_{21}x_1 + a_{22}x_2 + \cdots + a_{2n}x_n &= b_2 \\
\cdots\cdots\cdots\cdots\cdots\cdots\cdots \\
a_{n1}x_1 + a_{n2}x_2 + \cdots + a_{nn}x_n &= b_n
\end{aligned}
\tag{4.1}
$$

in which the coefficients a_{ij} and the b_i, $i, j = 1, 2, \ldots, n$, are each known constants, are simultaneous linear equations in the unknowns x_j. These equations are often written in the sigma-summation notation as

$$\sum_{j=1}^{n} a_{ij}x_j = b_i \qquad \text{for } i = 1, 2, \ldots, n \tag{4.1a}$$

Equations of this type occur frequently in practice, and one of the objectives of this chapter is to set down some systematic methods for solving for the unknowns x_j.

First, some definitions and rules pertaining to simultaneous linear equations are given.

1. If each b_i in the set (4.1) is equal to zero, the equations are homogeneous. If one or more of the b_i's in (4.1) is different from zero, the equations form a nonhomogeneous set.

2. If any one equation of (4.1) is a linear combination of any number of the others, the set is said to be linearly dependent. Conversely, if no equation in the set is a linear combination of the others, the system of equations is linearly independent.

3. If the set (4.1) is linearly independent and homogeneous, the only solution is that each x_j is zero. If the set (4.1) is linearly dependent and homogeneous, there is an infinite number of solutions.

4. If the set (4.1) is linearly independent, there is a unique solution for the unknowns x_j.

5. Any equation can be multiplied by a constant without changing the equality. Thus if the ith equation is multiplied by the constant c, one has

$$ca_{i1}x_1 + ca_{i2}x_2 + \cdots = cb_i \tag{4.2}$$

If this equation replaces the ith equation, the solution for the x_j's is unchanged.

6. Any equation or any equation modified as in (5) can be added or subtracted term by term to any other equation without changing the equality. For example, the kth equation can be replaced by adding to it the modified equation (4.2) to give

$$(a_{k1} + ca_{i1})x_1 + \cdots + (a_{kn} + ca_{in})x_n = b_k + cb_i \tag{4.3}$$

Neither operation (5) nor (6) changes the solution to the set. These observations will be used in constructing methods of solution for Eqs. (4.1).

Linear dependence of the set of equations in (4.1) depends upon the value of the determinant formed by the coefficients on the unknowns. This determinant is called the *characteristic determinant D.*

$$D = \begin{bmatrix} a_{11} & \cdots & a_{1j} & \cdots & a_{1n} \\ a_{21} & \cdots & a_{2j} & \cdots & a_{2n} \\ \cdots & \cdots & \cdots & \cdots & \cdots \\ a_{n1} & \cdots & a_{nj} & \cdots & a_{nn} \end{bmatrix} \tag{4.4}$$

If D is equal to zero, the equations are linearly dependent. If D is different from zero, the equations are linearly independent. A method for evaluating D is presented later.

4.2 Solution of Simultaneous Linear Equations

One of the simplest ways to describe a solution for Eqs. (4.1) is to use Cramer's rule, in which each x is expressed as a ratio of two determinants. By this rule the value of the first unknown is written as

$$x_1 = \frac{1}{D} \begin{bmatrix} b_1 & a_{12} & \cdots & a_{1n} \\ \cdots & \cdots & \cdots & \cdots \\ b_n & a_{n2} & \cdots & a_{nn} \end{bmatrix} \tag{4.5}$$

The formula for x_j is obtained by placing the b vector in the jth column of the determinant in Eq (4.5). The determinant in the numerator is the same as the denominator D, except that for x_j the jth column has been replaced by the constants $b_1, \ldots, b_n$.

The computation for x_1 requires the evaluation of two determinants. For each additional x computed by this method it is necessary to find the value of a different determinant in the numerator. For n unknowns there are $n + 1$ determinants, each n by n in size, to be evaluated. If n is large, this is a monumental task even for a high-speed machine. For this reason this method of calculation is almost never used. The form for the solution of x in Eq. (4.5) does give some useful information, however. It is used to verify some of the definitions and rules cited earlier.

If the characteristic determinant is zero, then Eq. (4.5) shows that unique values for x_j cannot exist. Conversely, when D is not zero, then x_j is uniquely defined. If the equations are homogeneous (all b's are zero), the numerator is zero for each x. For a set of linearly independent and homogeneous equations the only solution is that all unknowns be zero. If the equations are linearly independent and nonhomogeneous, then nonzero values for the unknowns are obtained. The test for linear independence is the nonvanishing of the characteristic determinant. This is an essential test, which is performed for any large system of equations and is usually made a part of the calculation routine.

A more efficient method of solution, which systematically eliminates unknowns from the set of equations, is now discussed.

4.3 An Elimination Method

The method of elimination presented here is a systematic machine-oriented procedure for efficiently solving large systems of simultaneous equations. The general idea of elimination methods is attributed to Gauss, and this method is called *Gaussian elimination.* The procedure is explained for a system of three equations in three unknowns and is then extended to an n by n system with a flow chart for machine computation. Consider the example

$$2x_1 + 4x_2 + 2x_3 = 16 \tag{4.6a}$$

$$2x_1 - x_2 - 2x_3 = -6 \tag{4.6b}$$

$$4x_1 + x_2 - 2x_3 = 0 \tag{4.6c}$$

By subtracting Eq. (4.6*a*) from Eq. (4.6*b*) one obtains

$$-5x_2 - 4x_3 = -22 \tag{4.7}$$

If Eq. (4.6*a*) is multiplied by 2 and then subtracted from (4.6*c*), one obtains

$$-7x_2 - 6x_3 = -32 \tag{4.8}$$

Equations (4.7) and (4.8) are a set of two equations in two unknowns from which another of the unknowns can be eliminated to give one equation in one unknown. Multiply Eq. (4.7) by $7/5$ and subtract the result from Eq. (4.8). The result is

$$-\tfrac{2}{5}x_3 = -\tfrac{6}{5}$$

from which $x_3 = 3$. By successively substituting back, first into (4.7) then into (4.6*a*), one finds

$$x_2 = 2$$

$$x_1 = 1$$

In the elimination calculations only the coefficients a_{ij} and the constants b_i have been used. The x's serve only to identify the particular unknown to which the coefficient applies. If the coefficients and constant terms are arranged so that there is no necessity for carrying the unknowns for identification, the latter can be dropped. Such is the case when the problem is represented by the following array of numbers:

$$\begin{bmatrix} 2 & 4 & 2 & 16 \\ 2 & -1 & -2 & -6 \\ 4 & 1 & -2 & 0 \end{bmatrix} \tag{4.9}$$

In this array, called a *matrix*, the first row represents the first equation, the second row the second equation, and so on. The first number in each row is the coefficient on the first unknown in the equation, the second number the coefficient on the second unknown, and so on. The last number in each row is the constant from that equation. This matrix can be visualized as the equations with the plus signs, unknowns, and equal signs removed. Thus, the second row of the matrix

$$2 \quad -1 \quad -2 \quad -6$$

represents the equation

$$2x_1 - 1x_2 - 2x_3 = -6$$

The elimination procedure used in the example is modified somewhat to make the computations more routine for machine computations. The first objective is to eliminate x_1 from the last two equations, i.e., to place zeros in the first-column position of rows 2 and 3. This is done in such a way that the value of the coefficients has no influence on the routine except in cases where they lead to division by zero.

The first operation on the matrix is with the first row, which is

$$2 \quad 4 \quad 2 \quad 16$$

Each element (number) in the row is divided by the first number to give

$$1 \quad 2 \quad 1 \quad 8$$

This sequence of numbers represents the first equation after it has been multiplied by ½. The equality of the equation is unchanged.

The second step in the elimination is to multiply this modified first row by the first coefficient in the second row and subtract the result from the second row. The second row, which is initially

$$2 \quad -1 \quad -2 \quad -6$$

now becomes

$$(2 - 2) \quad (-1 - 4) \quad (-2 - 2) \quad (-6 - 16)$$

or

$$0 \quad -5 \quad -4 \quad -22$$

A zero is placed in the first column of the second row by this operation. This is equivalent to eliminating the x_1 term between the first and second equations and is the same result as found previously in Eq. (4.7). The next operation is to place a zero in the first-column position of row 3, which is done by multiplying the modified row 1 by the leading coefficient of row 3 and then subtracting the resulting terms from row 3, which then becomes

$$(4 - 4) \quad (1 - 8) \quad (-2 - 4) \quad (0 - 32)$$

or

$$0 \quad -7 \quad -6 \quad -32$$

This set of numbers represents the equation obtained when x_1 is eliminated from the first and third equations and is exactly equivalent to Eq. (4.8) obtained earlier. The new matrix is now

$$\begin{bmatrix} 1 & 2 & 1 & 8 \\ 0 & -5 & -4 & -22 \\ 0 & -7 & -6 & -32 \end{bmatrix} \tag{4.10}$$

and represents the set of equations:

$$\begin{aligned} x_1 + 2x_2 + x_3 &= 8 \\ -5x_2 - 4x_3 &= -22 \\ -7x_2 - 6x_3 &= -32 \end{aligned} \tag{4.11}$$

The unknown x_1 has been eliminated from the last two equations, which now represent a system of two equations in two unknowns. The last two rows of the matrix represent these equations and can now be handled independently of the first row. The following operations are made on the last two rows:

1. Divide row 2 by its first element (here -5).
2. Multiply this new row 2 by the first element of row 3 (here -7) and subtract the result from row 3. The matrix then becomes

$$\begin{bmatrix} 1 & 2 & 1 & 8 \\ 0 & 1 & 4/5 & 22/5 \\ 0 & 0 & -2/5 & -6/5 \end{bmatrix} \tag{4.12}$$

The array (4.12) has the number 1 on the diagonal in all but the last row and has zeros in each position below the diagonal. It represents the following set of equations:

$$\begin{aligned} x_1 + 2x_2 + x_3 &= 8 \\ x_2 + \tfrac{4}{5}x_3 &= \tfrac{22}{5} \\ -\tfrac{2}{5}x_3 &= -\tfrac{6}{5} \end{aligned} \tag{4.13}$$

The value of x_3 is found from the third row of (4.12) or the third equation of (4.13).

$$x_3 = \frac{-6/5}{-2/5} = 3$$

When x_3 is known, the second row of (4.12) can be used to solve for x_2:

$$x_2 = 22/5 - 4/5 x_3 = 22/5 - (4/5)(3) = 2$$

and when x_3 and x_2 are known, the first row of (4.12) can be used to solve for x_1:

$$x_1 = 8 - 2x_2 - x_3 = 8 - (2)(2) - 3 = 1$$

The set of numbers

$$x_1 = 1 \qquad x_2 = 2 \qquad \text{and} \qquad x_3 = 3$$

is the solution to the equations represented by the array (4.9).

4.4 *Gaussian Elimination for n Equations*

The set of equations (4.1) can be represented by the matrix

$$\begin{bmatrix} a_{11} & a_{12} & \cdots & a_{1n} & b_1 \\ a_{21} & a_{22} & \cdots & a_{2n} & b_2 \\ \cdots & \cdots & \cdots & \cdots & \cdots \\ a_{n1} & a_{n2} & \cdots & a_{nn} & b_n \end{bmatrix} \tag{4.14}$$

In this array row 1 represents the first equation of the set, row 2, the second equation, and so forth. In column 1 are the coefficients on x_1, in column j the coefficients on x_j, and finally in the last column is the constant term in each equation. This is an n by $n + 1$ array of numbers and is called an n by $n + 1$ matrix.

In order to solve the set of equations represented by (4.14) the matrix is transformed into an array that has the number 1 in each position of the diagonal, i.e., $a_{kk} = 1$ for $k = 1, 2, \ldots, n$, and has zeros in all positions below the diagonal. A systematic method for accomplishing this follows the procedures used in the example and is outlined step by step below. In each operation some of the elements of the matrix will be changed. This is indicated by a superscript which tells the number of times the element has been modified. The original values in the matrix are identified by a zero superscript.

Considerable simplification can be made in the machine-calculation routine by changing the designation of the constant column to $a_{k,n+1}$ where k denotes the row. This in no way changes the value of the elements in the matrix. With the change in notation the initial matrix is given as

$$\begin{bmatrix} a_{11}^{(0)} & \cdots & a_{1n}^{(0)} & a_{1,n+1}^{(0)} \\ \cdots & \cdots & \cdots & \cdots \\ a_{n1}^{(0)} & \cdots & a_{nn}^{(0)} & a_{n,n+1}^{(0)} \end{bmatrix} \tag{4.15}$$

The first step in the elimination procedure is to place unity in the (1,1) position, which is done by dividing each term of the first row by $a_{11}^{(0)}$ ($a_{11}^{(0)} \neq 0$). Next the modified first row is used to eliminate the first column of each subsequent row. The operation on row 2 is to first multiply each term in the modified row 1 by the leading coefficient in row 2 and then subtract term by term from row 2. This places a zero in the first column of row 2. Each subsequent row is operated on by multiplying the modified row 1 by the leading coefficient for that row and then subtracting term by term from that row. When the set of operations is performed on each row, there is a zero in the first-column position of the last $n - 1$ rows and each coefficient in the matrix has been modified once in the elimination procedure, as denoted by the superscript (1) in

$$\begin{bmatrix} 1 & a_{12}^{(1)} & \cdots & a_{1,n+1}^{(1)} \\ 0 & a_{22}^{(1)} & \cdots & a_{2,n+1}^{(1)} \\ \cdots & \cdots & \cdots & \cdots \\ 0 & a_{n2}^{(1)} & \cdots & a_{n,n+1}^{(1)} \end{bmatrix} \tag{4.16}$$

The operation above has eliminated the first unknown from the last $n - 1$ equations. They are now a set of $n - 1$ equations in $n - 1$ unknowns. The same elimination procedure can be performed on this reduced set of equations so that unity will be placed in the position (2,2) and zeros in the second-column position for the last $n - 2$ rows. The modified matrix is given with superscripts to identify the number of times each entry has been modified. Clearly $a_{22}^{(1)}$ must not be zero.

$$\begin{bmatrix} 1 & a_{12}^{(1)} & a_{13}^{(1)} & \cdots & a_{1,n+1}^{(1)} \\ 0 & 1 & a_{23}^{(2)} & \cdots & a_{2,n+1}^{(2)} \\ 0 & 0 & a_{33}^{(2)} & \cdots & a_{3,n+1}^{(2)} \\ \cdots & \cdots & \cdots & \cdots & \cdots \\ 0 & 0 & a_{n3}^{(2)} & \cdots & a_{n,n+1}^{(2)} \end{bmatrix} \tag{4.17}$$

Now the last $n - 2$ rows represent a set of $n - 2$ equations in $n - 2$ unknowns. They in turn are operated on by the elimination method to give a matrix which contains a set of $n - 3$ equations in $n - 3$ unknowns. This comprises the third modification of the elements of the determinant. Repeating the elimination procedure each time on the reduced matrix will give, after $n - 1$ times, the matrix (4.18). In the kth divide operation it is always assumed that $a_{kk}^{(k-1)}$ is different from zero. Some modification of the method is indicated if this is not the case and will be discussed later in this chapter.

$$\begin{bmatrix} 1 & a_{12}^{(1)} & \cdots & a_{1,n-1}^{(1)} & a_{1,n}^{(1)} & a_{1,n+1}^{(1)} \\ 0 & 1 & \cdots & a_{2,n-1}^{(2)} & a_{2,n}^{(2)} & a_{2,n+1}^{(2)} \\ \cdots & \cdots & \cdots & \cdots & \cdots & \cdots \\ 0 & 0 & \cdots & 1 & a_{n-1,n}^{(n-1)} & a_{n-1,n+1}^{(n-1)} \\ 0 & 0 & \cdots & 0 & a_{n,n}^{(n-1)} & a_{n,n+1}^{(n-1)} \end{bmatrix} \tag{4.18}$$

The equation represented by the last row of (4.18) is

$$a_{n,n}^{(n-1)} x_n = a_{n,n+1}^{(n-1)} \tag{4.19}$$

which is readily solved for x_n. When x_n is known, the $(n - 1)$st row is used to solve for x_{n-1}.

$$x_{n-1} = a_{n-1,n+1}^{(n-1)} - a_{n-1,n}^{(n-1)} x_n$$

By successive substitution of known values of x all the unknowns can be found. The kth unknown is found by using the kth row of (4.18):

$$x_k + \sum_{j=k+1}^{n} a_{kj}^{(k)} x_j = a_{k,n+1}^{(k)} \tag{4.20}$$

Solving Eq. (4.20) for x_k gives

$$x_k = a_{k,n+1}^{(k)} - \sum_{j=k+1}^{n} a_{kj}^{(k)} x_j \tag{4.21}$$

Applying this formula for $k = n - 1, n - 2, \ldots, 2, 1$ will give the complete solution of the system (4.15).

4.5 Formal Steps in the Gaussian Elimination Method

The steps in the elimination method are outlined in the order in which they are done in machine computation. Each entry in the matrix is assumed to represent the current value for that entry. Since the final computations for the unknowns are made in terms of the last entry in the matrix, it is unnecessary to keep track of the original values or any intermediate values. In outlining the steps the destructive read-in property of the memory of a computing machine is utilized. Thus each time an entry in the matrix is modified, the new value replaces the old value in storage, and so the superscript notation can be dropped. It is assumed that a_{kk} is not equal to zero when the divide operation is made. Modification in the routine for $a_{kk} = 0$ is considered later.

1. Divide row 1 by a_{11}.

2. Multiply row 1 by a_{k1} and subtract from row k, first for $k = 2$, then for $k = 3$, etc., through $k = n$. This places a zero in the first column of rows 2 through n.

3. Divide row 2 by a_{22}.

4. Multiply row 2 by a_{k2} and subtract from row k for $k = 3, 4, \ldots, n$. This places a zero in the second column for rows 3 through n.

5. Repeat steps 3 and 4, each time increasing the row number until the row number is equal to $n - 1$.

When these steps are completed, the matrix has the form of (4.18). Calculation of the unknowns is now made by successive substitution using Eqs. (4.19) and (4.21). A flow chart for this problem is shown in Fig. 4.1. A detailed explanation follows.

1. The number of equations n and the matrix A are read into locations N and A. The matrix elements are designated A(I,J) for $J = 1, 2, \ldots, n + 1$ and $I = 1, 2, \ldots, n$. The element A(I, N + 1) is the constant term in the ith equation.

2. This iteration statement sets the control for the successive elimination of the first $n - 1$ unknowns. The operations to eliminate the unknown x_k from the last $n - k$ equations, as described above, are performed under the control of this statement.

3. The diagonal element of row k is placed into location B. In this way, its value will not be destroyed in statement 5. B is used as a temporary storage location, and its contents will be saved only as long as needed.

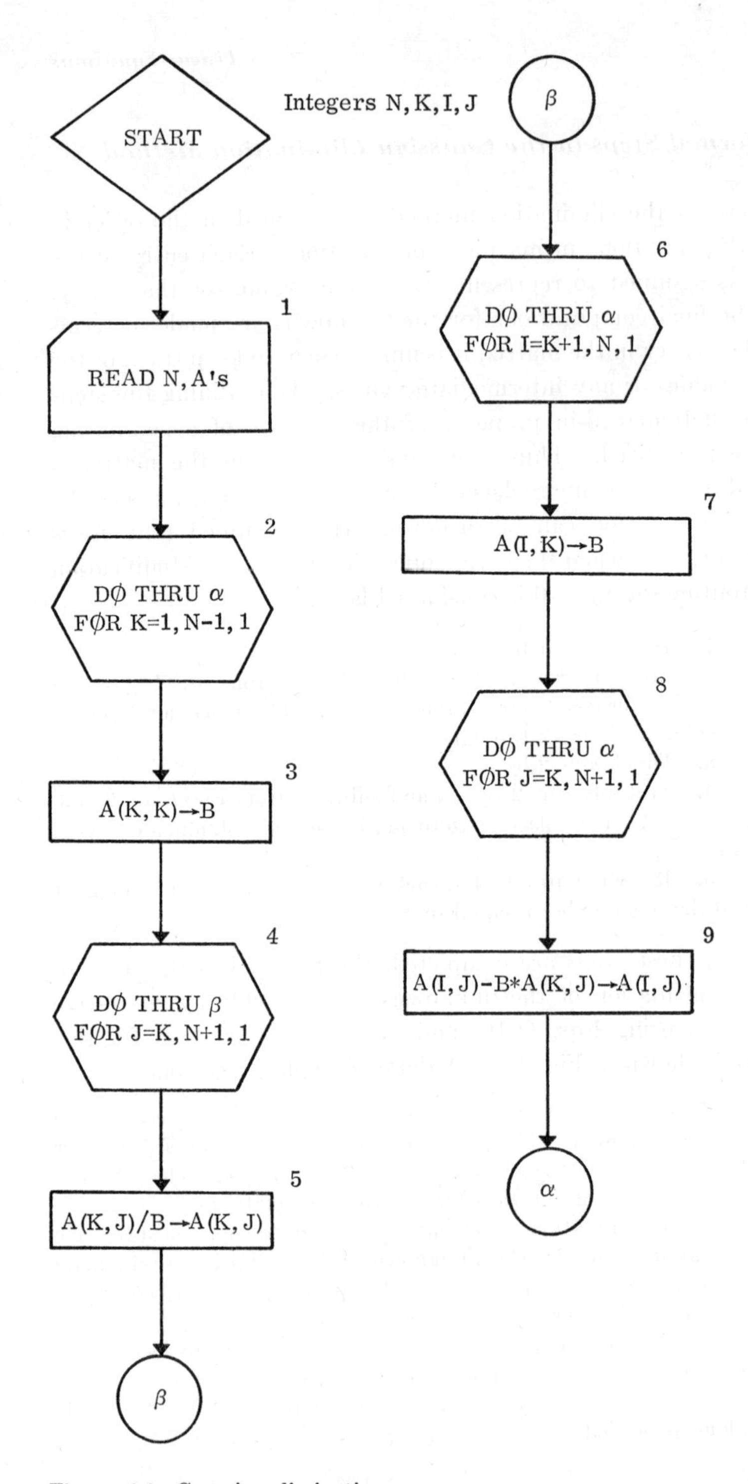

Figure 4.1 Gaussian elimination.

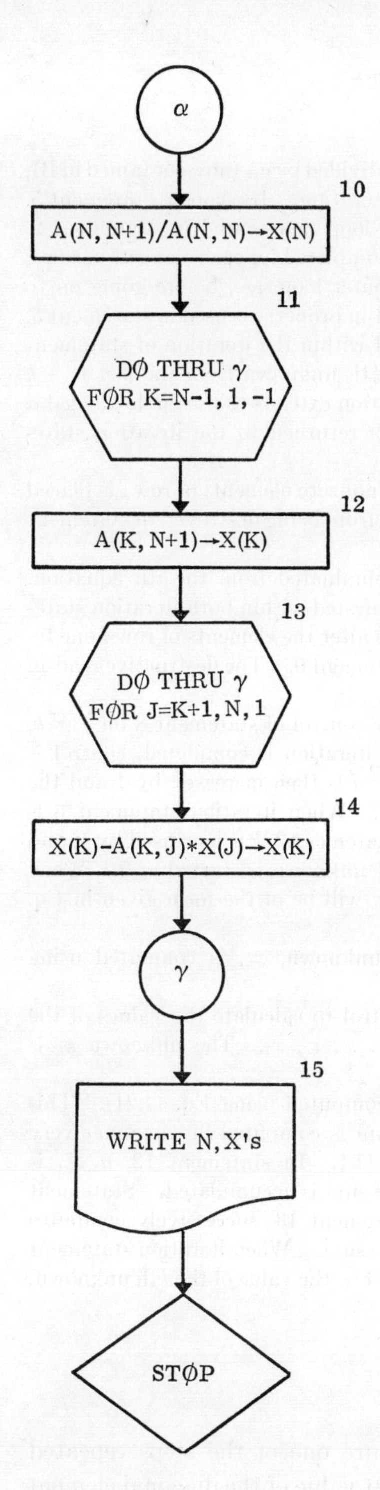

Figure 4.1 *(Continued)*

4, 5. Each number in row k is divided by a_{kk} (now contained in B), $a_{kk} \neq 0$. Statement 4 is a nested iteration. Its scope is statement 5, and it is entirely within the outer loop represented by statement 2. The operations in statement 5 are completed for each element in row k before control is returned to the outer loop, i.e., before going on to statement 6. The destructive read-in property is used in statement 5.

6. This iteration, also nested within the iteration of statement 2, is the control to eliminate the kth unknown from the last $n - k$ equations. The scope of this iteration extends to the circle labeled α and is completed before control is returned to the iteration statement 2.

7. The kth element (the first nonzero element) of row i is placed into location B to prevent its value from being destroyed in statement 9.

8, 9. The kth unknown is eliminated from the ith equation. Statement 8 is an iteration that is nested within both iteration statements 2 and 6. It is the control to alter the elements of row i one by one according to the arithmetic statement 9. The destructive read-in property is used in statement 9.

Statement 9 is executed under control of statement 8 for $j = k, k + 1, \ldots, n + 1$. When this iteration is completed, control is returned to iteration statement 6. i is then increased by 1 and the statements repeated starting at 7. When iteration statement 6 is completed, control is returned to statement 2, k is increased by 1, and the process of eliminating the next unknown is started at 3. When statement 2 is satisfied, the matrix will be of the form given in Eq. (4.18).

10. The value of the last unknown, x_n, is computed using Eq. (4.19) and stored in X(N).

11. This iteration is the control to calculate the values of the unknowns in the order $x_{n-1}, x_{n-2}, \ldots, x_1$. The unknown x_k is stored in X(K).

12–14. The value for x_k is computed using Eq. (4.21). This involves a sum of terms. This sum is computed in a manner very similar to that used in Example 2.1. In statement 12, $a_{k,n+1}$ is placed in location X(K), where the sum is accumulated. Statement 14, under control of iteration statement 13, successively computes each term and subtracts it from the sum. When iteration statement 13 is satisfied, the value in X(K) will be the value of the kth unknown.

4.6 The Problem of Zeros

In the Gaussian elimination procedure one of the steps repeated for each row is to divide by the current value of the diagonal element of that row. Should this number be zero, this operation is not admissible. If the element is quite small compared with other elements in the row, the division leads to very large numbers that can invalidate the remainder of the calculations. When this problem arises, it can be resolved by simply interchanging this row with any row below it for which the element in this particular column is not zero (or very small). Clearly this is a valid operation, since the order in which a set of equations is written down in no way influences the solution.

To ensure that division by a relatively small number will not occur, it is necessary to test the diagonal element on each row immediately before the division step. For the kth row this would be the current value for the kth element, viz., a_{kk}. The absolute value of this element is compared to some arbitrary quantity, and if it is greater than this number, the computations proceed as before. If it is less, a search is made down this column from the kth row until an element is found that is large enough to satisfy the given criterion. The row containing this element is then interchanged with the kth row, and the computations proceed as before. If no element can be found which is large enough, a unique solution cannot be found, and it is probable that the equations are linearly dependent; if no element can be found that is different from zero, the equations are surely linearly dependent.

For the greatest possible accuracy each column can be searched for the largest (in absolute value) candidate and then that row interchanged with the row under consideration. This has the effect of keeping all numbers in the matrix near to the same size.

The modification of the flow chart in Fig. 4.1 to test the diagonal element before division is left as an exercise. (See Fig. 4.7 for an example of row interchange.)

4.7 *Determinants*

An nth-order determinant is a square array of n^2 elements. An n by n determinant D with elements a_{ij} is written

$$D = \begin{vmatrix} a_{11} & a_{12} & \cdots & a_{1n} \\ a_{21} & a_{22} & \cdots & a_{2n} \\ \cdots & \cdots & \cdots & \cdots \\ a_{n1} & a_{n2} & \cdots & a_{nn} \end{vmatrix} \tag{4.22}$$

The elements a_{ij} may be any algebraic quantities or functions. In the discussion which follows attention is focused only on evaluating the determinant when the numerical value for each element is known, each element thus being regarded as a number.

The value of a determinant is defined as the sum of the product of each element of any row or column and its cofactor. A cofactor is the minor of an element with the appropriate algebraic sign. When this definition is used, the value of D in Eq. (4.22) can be written

$$D = \sum_{i=1}^{n} a_{ij}A_{ij} = \sum_{i=1}^{n} a_{ij}(-1)^{i+j}M_{ij} \tag{4.23}$$

where M_{ij} is the determinant formed by striking out the row and column containing a_{ij}. M_{ij} is called the *minor* of a_{ij}, and A_{ij} is the cofactor of a_{ij}. The sign of the cofactor is based on the choice of the row and column:

$$A_{ij} = (-1)^{i+j}M_{ij} \tag{4.24}$$

The value of the determinant can as well be represented by summing on the column index:

$$D = \sum_{j=1}^{n} a_{ij}A_{ij} \tag{4.25}$$

where again Eq. (4.24) holds.

This method for evaluation of a determinant is the well-known method of expanding the determinant by minors. The determinants M_{ij} are $n - 1$ by $n - 1$ determinants, and each of these can in turn be expanded by minors to give $n - 2$ by $n - 2$ determinants. Clearly this can be carried on until each minor is a 2 by 2 determinant that can be calculated by the rule for expanding a second-order determinant.

$$\begin{vmatrix} b_{11} & b_{12} \\ b_{21} & b_{22} \end{vmatrix} = b_{11}b_{22} - b_{12}b_{21} \tag{4.26}$$

The calculations suggested above do not give an efficient method for finding the value of a determinant. A much faster and simpler technique based on the Gaussian elimination procedure of the previous section is normally used. It depends upon some properties of determinants which are listed below.

1. If any row or column of a determinant is multiplied by a constant, the value of the determinant will be multiplied by the same constant.

2. If any two rows (or columns) of a determinant are interchanged, the sign of the determinant is changed. In general, if there are k interchanges of rows and/or columns, the sign of the determinant will be changed by $(-1)^k$.

3. If any row (or column) of a determinant is equal term by term to any other row (or column), the determinant is equal to zero.

4. If any multiple of one row is added to (or subtracted from) another row or if any multiple of one column is added to (or subtracted from) another column, the value of the determinant is not changed.

5. If any row (column) of a determinant is a linear combination of any row or rows (column or columns), the determinant is equal to zero.

These statements can be proved by using the definition given in Eq. (4.23) or (4.25). They are used in adapting the Gaussian elimination method for evaluating a determinant.

4.8 *Evaluating a Determinant by Gaussian Elimination*

The method of Gaussian elimination for evaluating a determinant is illustrated with the following example:

$$D = \begin{vmatrix} 2 & 4 & 2 \\ 2 & -1 & -2 \\ 4 & 1 & -2 \end{vmatrix} \tag{4.27}$$

This is the characteristic determinant of the system of equations represented by the matrix in Eq. (4.9).

The first step in the procedure to evaluate D is to divide the first row by 2 which is the leading element of row 2 and to multiply the determinant by this same number. Because of property 1 above, this does not change the value of the determinant.

$$D = 2\begin{vmatrix} 1 & 2 & 1 \\ 2 & -1 & -2 \\ 4 & 1 & -2 \end{vmatrix}$$

The first elements of rows 2 and 3 are eliminated by the Gaussian elimination process. From property 4 of determinants this operation will not alter the value of the determinant.

$$D = 2 \begin{vmatrix} 1 & 2 & 1 \\ 0 & -5 & -4 \\ 0 & -7 & -6 \end{vmatrix} \tag{4.28}$$

A comparison reveals that, aside from the factor 2, this determinant is the characteristic determinant of the equations represented by the matrix in Eq. (4.10). The only difference in the computations for obtaining the two arrays is that the divisor becomes a factor in the value of the determinant.

If the determinant in (4.28) is expanded by minors on its first column, the result is

$$D = 2 \begin{vmatrix} -5 & -4 \\ -7 & -6 \end{vmatrix}$$

Hence, attention can be focused on the last two rows of (4.28). The following operations are made on these last two rows: (1) Divide row 2 by its first element (here -5) and multiply the determinant by the same number; (2) multiply the modified row 2 by the first element in row 3 (here -7) and subtract the result from row 3. The determinant then becomes

$$D = -10 \begin{vmatrix} 1 & 2 & 1 \\ 0 & 1 & 4/5 \\ 0 & 0 & -2/5 \end{vmatrix} \tag{4.29}$$

This determinant is readily expanded by minors to give

$$D = (-10)(-2/5) = 4$$

The only differences between this method of evaluating determinants and the Gaussian elimination for solving equations are that the $(n + 1)$st column of the matrix is missing, and each time a row is divided by its leading element, the determinant is multiplied by that same number. The value of the determinant is the product of all the divisors and the last element of the array.

A flow chart to evaluate a determinant with elements a_{ij}, i, $j = 1, 2, \ldots, n$, by the Gaussian elimination method is given

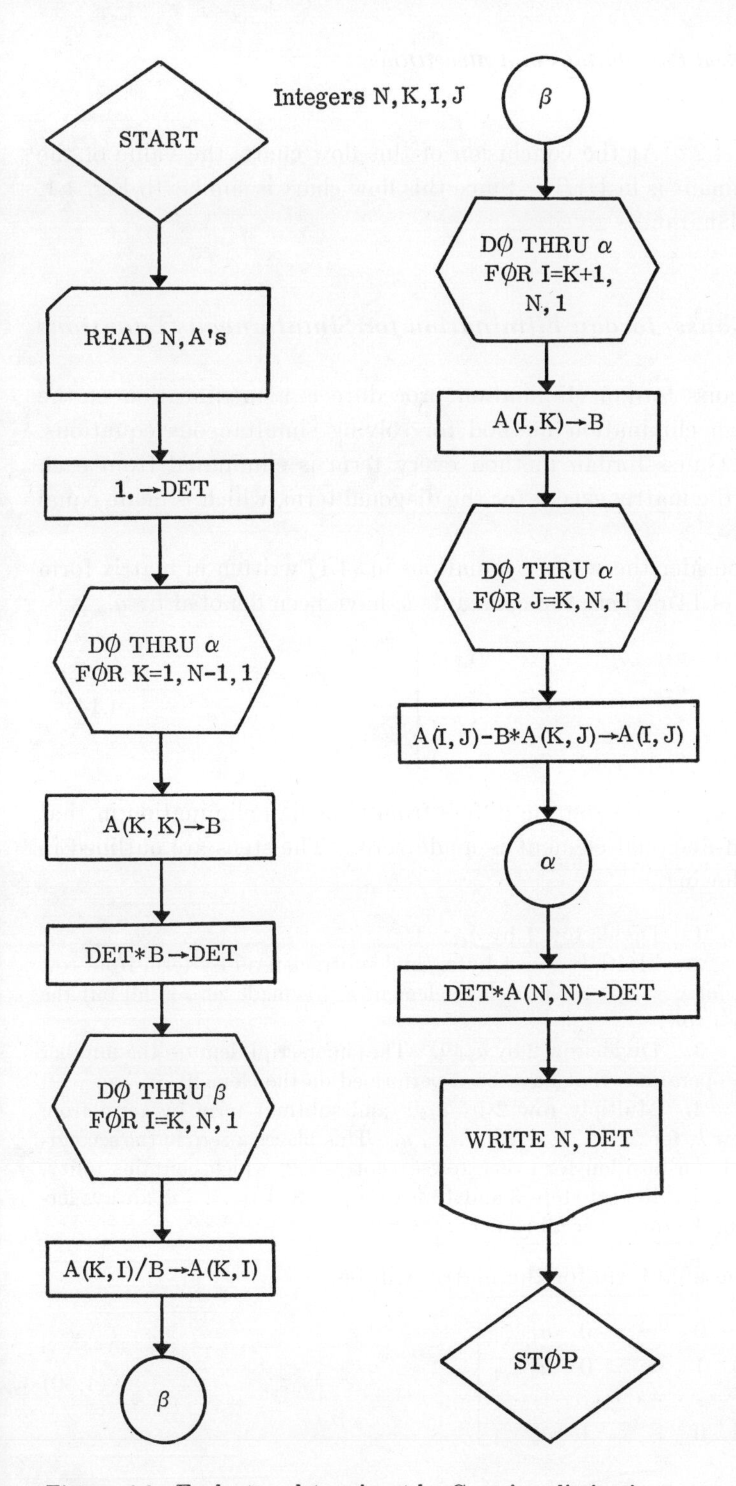

Figure 4.2 Evaluate a determinant by Gaussian elimination.

in Fig. 4.2. At the conclusion of this flow chart, the value of the determinant is in DET. Since this flow chart is similar to Fig. 4.1, no explanation is given.

4.9 *Gauss-Jordan Elimination for Simultaneous Equations*

The Gauss-Jordan elimination procedure is a modification of the Gaussian elimination method for solving simultaneous equations. In the Gauss-Jordan method every term is eliminated from each row of the matrix except for the diagonal term, which is made equal to unity.

Consider the set of n equations in (4.1) written in matrix form in Eq. (4.14), where the constants b_i have been denoted by $a_{i,n+1}$.

$$\begin{bmatrix} a_{11} & a_{12} & \cdots & a_{1n} & a_{1,n+1} \\ a_{21} & a_{22} & \cdots & a_{2n} & a_{2,n+1} \\ \cdot & \cdot & \cdot & \cdot & \cdot \\ a_{n1} & a_{n2} & \cdots & a_{nn} & a_{n,n+1} \end{bmatrix} \tag{4.14}$$

The computation routine differs from Gaussian elimination in that each off-diagonal element is made zero. The steps are outlined in the following.

1. Divide row 1 by a_{11}.
2. Multiply row 1 by a_{k1} and subtract term by term from row k, for $k = 2, \ldots, n$. The element $a_{k,1}$ is made zero in all but the first row.
3. Divide row 2 by $a_{22}^{(1)}$. The superscript denotes the number of operations that have been performed on the element.
4. Multiply row 2 by $a_{k2}^{(1)}$ and subtract term by term from row k, for $k = 1, 3, 4, 5, \ldots, n$. This places a zero in the second-column position for every row except row 2, which contains unity.
5. Repeat steps 3 and 4 for rows $j = 3, 4, \ldots, n$, always letting k range over all rows except row j.

The final form for the matrix will be

$$\begin{bmatrix} 1 & 0 & \cdots & 0 & a_{1,n+1}^{(n)} \\ 0 & 1 & \cdots & 0 & a_{2,n+1}^{(n)} \\ \cdot & \cdot & \cdot & \cdot & \cdot \\ 0 & 0 & & 1 & a_{n,n+1}^{(n)} \end{bmatrix} \tag{4.30}$$

Each equation represented by this matrix is of the form

$$x_k = a_{k,n+1}^{(n)}$$

for $k = 1, 2, \ldots, n$. Thus, the solution for the unknowns is in the last column of the matrix.

4.10 Gauss-Seidel Iteration

The Gaussian elimination and Gauss-Jordan elimination for solving simultaneous equations belong to a class of methods called *direct methods.* They are characterized by the fact that after a finite number of computations and in the absence of round-off error they give exact results.

A second class comprises methods which depend upon successively improving an approximation to the solution and which are called *iterative methods.* They do not give an exact solution to the problem but get closer and closer to the solution as the number of computations is increased indefinitely, provided the process converges. They may give better results than the direct methods if round-off errors are significant.

The Gauss-Seidel iteration method is a procedure for successively improving an initial estimate for the unknowns in Eq. (4.1). The algorithm for successive improvement of the solution is obtained from Eq. (4.1) by solving the ith equation for x_i.

$$x_i = \frac{b_i - \sum_{\substack{j=1 \\ j \neq i}}^{n} a_{ij}x_j}{a_{ii}} \qquad i = 1, 2, \ldots, n \tag{4.31}$$

The variable j in the sum ranges over all values from 1 to n inclusive except $j = i$.

Let an initial estimate to the solution be represented by $x_j^{(0)}$, $j = 1, 2, \ldots, n$. The superscript will be used to denote the number of iterations that have been made. An improved value for x_1 is obtained by setting $i = 1$ in Eq. (4.31) and using the initial estimates to the unknowns to evaluate the right-hand side. The improved value for x_1 is designated $x_1^{(1)}$ where, as noted above, the superscript denotes the number of iterations.

$$x_1^{(1)} = \frac{b_1 - \sum_{j=2}^{n} a_{ij}x_j^{(0)}}{a_{11}}$$

This value, along with the initial guess $x_j^{(0)}$, $j = 3, 4, \ldots, n$, is then used in Eq. (4.31) with $i = 2$ to give an improved value for x_2.

$$x_2^{(1)} = \frac{b_2 - a_{21}x_1^{(1)} - \sum_{j=3}^{n} a_{2j}x_j^{(0)}}{a_{22}}$$

In a similar way improved values for all the unknowns are obtained. The general formula used for $x_i^{(1)}$ is

$$x_i^{(1)} = \frac{b_i - \sum_{j=1}^{i-1} a_{ij}x_j^{(1)} - \sum_{j=i+1}^{n} a_{ij}x_j^{(0)}}{a_{ii}} \tag{4.32}$$

for $i = 1, 2, \ldots, n$. These improved values $x_i^{(1)}$ are then used in place of $x_i^{(0)}$ in a second iteration to obtain $x_i^{(2)}$. The $(p + 1)$st iteration is made with the following set of equations:

$$x_i^{(p+1)} = \frac{b_i - \sum_{j=1}^{i-1} a_{ij}x_j^{(p+1)} - \sum_{j=i+1}^{n} a_{ij}x_j^{(p)}}{a_{ii}} \qquad i = 1, 2, \ldots, n \tag{4.33}$$

which are used to improve the approximation to the solution until suitable values for the unknowns are obtained or until it is established that further iterations will not improve the solution.

The test to determine the suitability of a result is made on the basis of the change in the solution in successive iterations. If

$$\frac{x_i^{(p+1)} - x_i^{(p)}}{x_i^{(p)}} < \epsilon \qquad \text{for } i = 1, 2, \ldots, n \tag{4.34}$$

where ϵ is an acceptable error criterion, the solution $x_i^{(p+1)}$ is a satisfactory result.

In the computations for the Gauss-Seidel method the new approximation to each unknown replaces the old value as soon as it is available. Thus there is no necessity of identifying the order of the iterations, and the superscript notation is dropped. This fact simplifies the operations in a computer program.

Figure 4.3 is a flow chart for Gauss-Seidel iteration. In the following explanation, the obvious steps are not covered. Convergence of the iterations is assumed in the flow chart.

1. The data which define the set of equations, an error test value, and an initial guess for the solution are read into the machine. ERR contains the value that is used in testing the solution for the desired accuracy.

2. At the end of each iteration, DMAX contains the largest change made in an unknown in that iteration. In statement 2, DMAX is initialized to zero.

3. This is the control to compute the improved values for the unknowns.

4–8. Equation (4.33) is used to compute the improved value for x_i. This involves a summing procedure and a division by a_{ii}. After completing statement 8, the value in XNEW is the improved value for x_i.

9, 10. The magnitude of the relative change in x_i is compared with DMAX and, if larger, is placed in DMAX. When all the unknowns have been improved, DMAX contains the largest relative change in any unknown.

11. After the comparison the improved value is placed into X(I) to be used in future computations. This is the end of the loop controlled by statement 3. Statements 4 through 11 are repeated for each of the unknowns.

12, 13. The largest change, DMAX, is compared to ERR. If this change is larger than ERR, control is returned to perform another iteration. If the change is smaller than ERR, the solution is written out.

4.11 Convergence of Gauss-Seidel Iteration

If the Gauss-Seidel iterations converge, they converge to the correct solution. The best way to determine convergence on a high-speed digital computer is to compare the results of successive iterations. Practically, if the value for DMAX, which is the largest relative

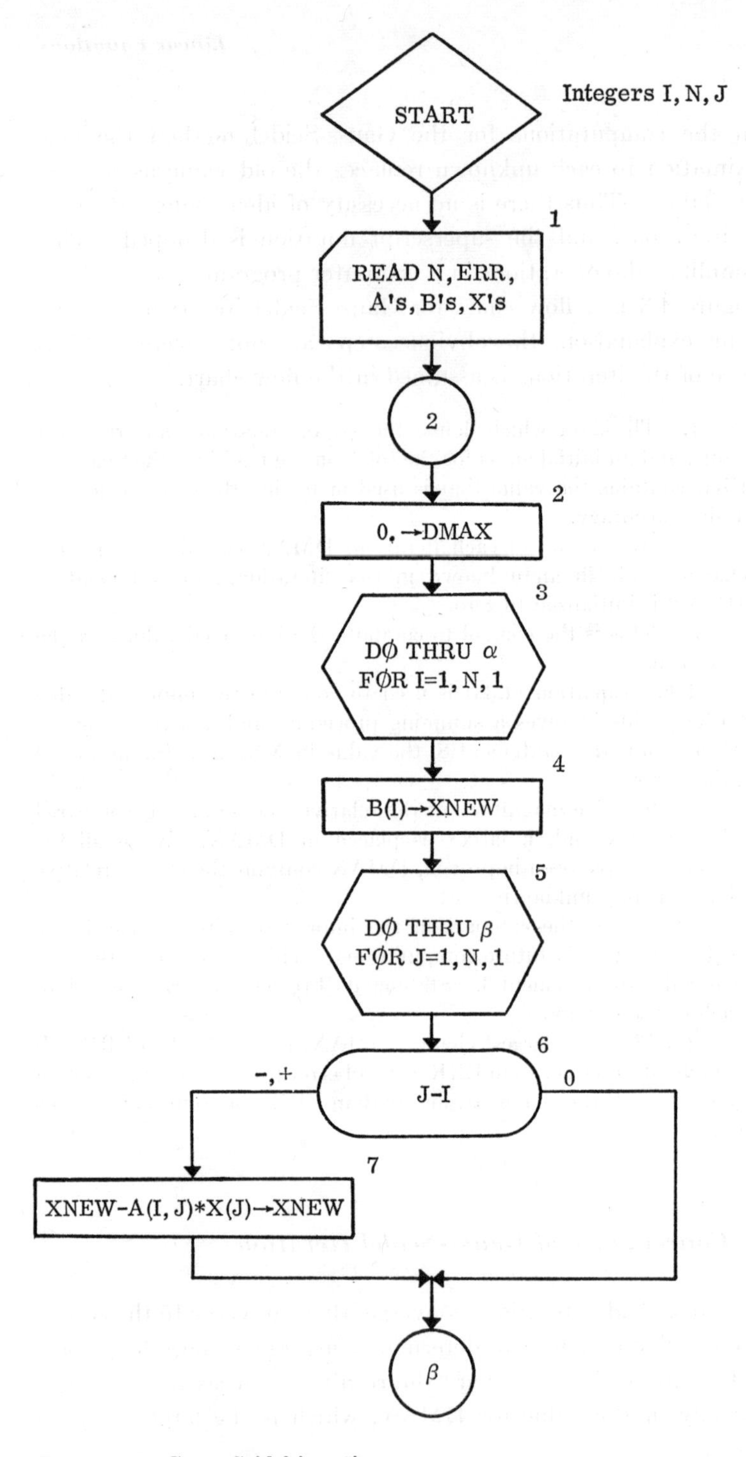

Figure 4.3 Gauss-Seidel iteration.

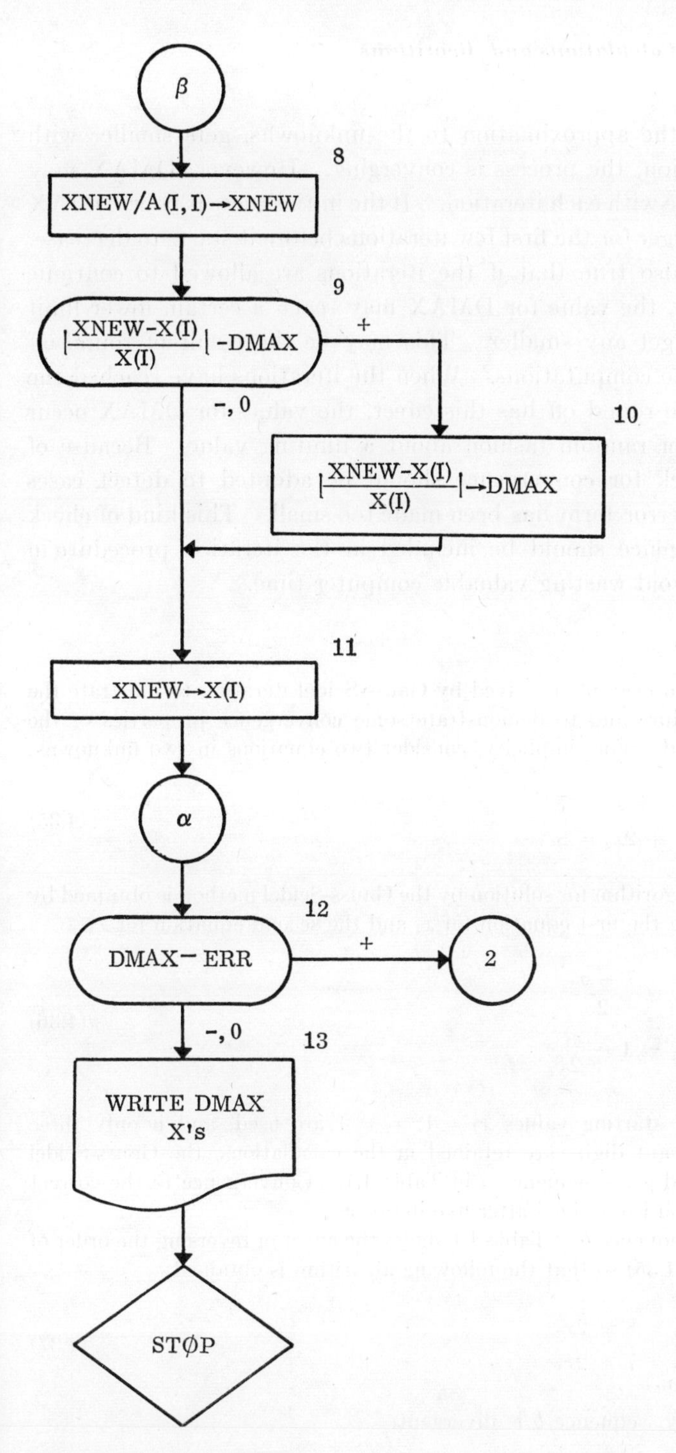

Figure 4.3 *(Continued)*

change in the approximation to the unknowns, gets smaller with each iteration, the process is converging. However, DMAX may not decrease with each iteration. If the initial guess is poor, DMAX may get larger for the first few iterations before it starts to decrease.

It is also true that if the iterations are allowed to continue indefinitely, the value for DMAX may reach a certain lower limit and never get any smaller. This may be the result of round-off errors in the computations. When the iterations have reached the point where round off has this effect, the values for DMAX occur in a kind of random fashion about a limiting value. Because of this a check for convergence should be adopted to detect cases where the error term has been made too small. This kind of check for convergence should be included in the iteration procedure in order to avoid wasting valuable computer time.

Example 4.1

An example is solved by Gauss-Seidel iteration to illustrate the procedure and to demonstrate some convergence properties of the method. For simplicity, consider two equations in two unknowns:

$$\begin{aligned} 2x_1 + x_2 &= 7 \\ x_1 + 2x_2 &= 8 \end{aligned} \tag{4.35}$$

The algorithm for solution by the Gauss-Seidel method is obtained by solving the first equation for x_1 and the second equation for x_2.

$$\begin{aligned} x_1 &= \frac{7 - x_2}{2} \\ x_2 &= 4 - \frac{x_1}{2} \end{aligned} \tag{4.36}$$

If starting values $x_1 = 1$, $x_2 = 1$ are used, and if only three significant digits are retained in the calculations, the Gauss-Seidel method gives sequence a in Table 4.1. Convergence to the correct solution is obtained after five iterations.

Sequence b of Table 4.1 shows the effect of reversing the order of Eqs. (4.35) so that the following algorithm is obtained:

$$\begin{aligned} x_1 &= 8 - 2x_2 \\ x_2 &= 7 - 2x_1 \end{aligned} \tag{4.37}$$

Clearly, sequence b is divergent.

Table 4.1

k	0	1	2	3	4	5
			Sequence a			
x_1	1	3.00	2.25	2.06	2.01	2.00
x_2	1	2.50	2.88	2.97	3.00	3.00
			Sequence b			
x_1	1	6	18	64		
x_2	1	−5	−29	−121		

The difference in the two sets of iteration equations is the size of the diagonal element a_{ii} in the coefficient matrix. If the diagonal element is significantly larger than the other elements in the row, the iteration generally will converge; otherwise, it may not. The best test for divergence is a comparison of the difference between successive approximations for the unknowns. If after a number of iterations the difference is increasing, the process is probably divergent. In any event, further investigation is in order before proceeding. Reference is made to the error discussion in Chap. 3, which is relevant for any iterative scheme based on improving an initial approximation.

4.12 Comparison of Methods for Solving Simultaneous Equations

Two distinctly different methods have been presented for solving simultaneous equations. The first is a direct method of elimination, which in the absence of round-off errors gives exact results. The second is an iteration procedure, which can often be applied so that it converges to the solution after many repetitions. An evaluation of the methods for use in machine computations is based primarily on the time required to obtain a satisfactory result.

Of the two direct methods Gaussian elimination is preferred because it involves fewer computations and therefore requires less time. The saving in time when using Gaussian elimination may be as high as 20 to 30 percent of the time required for the Gauss-Jordan elimination.

The choice of Gaussian elimination or Gauss-Seidel iteration for solving simultaneous equations is more difficult to make. The number of arithmetic operations in one iteration of the Gauss-Seidel method is approximately $2n^2 + n$, where n is the number of equations. The number of arithmetic operations for Gaussian elimination is roughly $n^3/3 + n^2$. Since several iterations are normally made for Gauss-Seidel, the elimination is easily faster when n is small. When n is equal to 100, approximately 16 iterations could be made with the same number of machine operations as the Gaussian elimination would require. Still the Gaussian elimination might be preferred because there is no assurance that 16 iterations of Gauss-Seidel will give suitable results.

If n is large, the problem usually has some special characteristics that will be incorporated into the solution. Generally many of the coefficients are zero, and those which are not are in a band to either side of the main diagonal of the coefficient matrix. For such problems the elimination method is still found to be superior for n as large as 500. This is not to say that the iteration method should not be used. It does in fact compete favorably with the elimination scheme where the term on the main diagonal is dominant. It should also be noted that there are means available whereby the convergence can be made much more general and much faster than for the Gauss-Seidel method. These techniques[9,12] make the iteration method much more attractive.

4.13 Introduction to Matrix Algebra

The purpose of this section is to define some of the basic matrix operations and to present some elementary applications of matrices.

A matrix is a rectangular array of numbers or functions called *elements*. The size of a matrix is measured by the number of rows and the number of columns in the array. Thus, a matrix with m rows and n columns is an m by n matrix and contains m times n elements. The elements of a matrix are designated by a small letter with two subscripts, of which the first denotes the row of the element and the second the column. Thus, the element a_{ij} of the matrix A is that number in the ith row and jth column of A.

The m by n matrix A with elements a_{ij} is written

$$A = [a_{ij}] = \begin{bmatrix} a_{11} & a_{12} & \cdots & a_{1n} \\ a_{21} & a_{22} & \cdots & a_{2n} \\ \cdots & \cdots & \cdots & \cdots \\ a_{m1} & a_{m2} & \cdots & a_{mn} \end{bmatrix} \tag{4.38}$$

Square brackets are used to indicate that the elements form a matrix. Some definitions and properties of matrices are listed below.

1. Matrix A is equal to matrix B if and only if the corresponding elements of A and B are equal. $A = B$ if and only if $a_{ij} = b_{ij}$ for each i and j. To be equal the two matrices A and B must be the same size, i.e., have the same number of rows and the same number of columns.

2. The null, or zero, matrix is a matrix in which every element is zero.

3. A square matrix has the same number of columns as it has rows.

4. The unit, or identity, matrix is a square matrix with elements δ_{ij} and is designated I.

$$I = [\delta_{ij}] \qquad \delta_{ij} = 0 \qquad \text{if } i \neq j,\ \delta_{ii} = 1$$

The unit matrix has unity in each element of the main diagonal and zero elsewhere.

5. A vector is a matrix with only one column or one row. The elements of a vector are usually written with only one subscript. Thus the column vector X with elements x_j is written

$$X = [x_j] = \begin{bmatrix} x_1 \\ x_2 \\ \cdot \\ \cdot \\ \cdot \\ x_m \end{bmatrix} \tag{4.39}$$

If the vector X has m components, it is called an m-dimensional vector.

6. The transpose of a matrix is the matrix with the columns and rows interchanged. Thus, the transpose of A, written A^t, is

defined as

$$A^t = \begin{bmatrix} a_{11} & a_{12} & \cdots & a_{1n} \\ a_{21} & a_{22} & \cdots & a_{2n} \\ \cdots & \cdots & \cdots & \cdots \\ a_{m1} & a_{m2} & \cdots & a_{mn} \end{bmatrix}^t = \begin{bmatrix} a_{11} & a_{21} & \cdots & a_{m1} \\ a_{12} & a_{22} & \cdots & a_{m2} \\ \cdots & \cdots & \cdots & \cdots \\ a_{1n} & a_{2n} & \cdots & a_{mn} \end{bmatrix} \tag{4.40}$$

The transpose of an m by n matrix is an n by m matrix.

7. A symmetric matrix is a square matrix that is equal to its transpose. If $A = A^t$, then A is a symmetric matrix, and $a_{ji} = a_{ij}$ for each i and j.

8. The length of a vector is the square root of the sum of the squares of the elements. The length of the vector $X = [x_j]$ is designated $\|X\|$.

$$\|X\| = \sqrt{\sum_{j=1}^{m} x_j^2} = \sqrt{X^tX} \tag{4.41}$$

Equation (4.41) follows from 5 and 6.

The algebraic operations of addition, subtraction, and multiplication for matrices are different from the corresponding operations with scalars (numbers). They are defined below and form the basis for the algebra of matrices.

1. The sum (or difference) of two matrices A and B is the matrix C, whose elements are the sum (or difference) of the corresponding elements of A and B: $A + B = C$ means $a_{ij} + b_{ij} = c_{ij}$ for each i and j. Clearly the matrices A, B, and C must all be the same size for either addition or subtraction.

2. The product of a scalar k and a matrix A is the matrix B, whose elements are the product of k and the corresponding elements of A: $kA = B$ means $ka_{ij} = b_{ij}$ for each i and j.

3. The product of a matrix A by a matrix B, written as AB, is defined when the number of columns in A is equal to the number of rows in B. The result is a matrix C which has the same number of rows as A and the same number of columns as B. Let A be a matrix with m rows and n columns and B be a matrix with n rows and l columns. The product $AB = C$ is defined and is a matrix with m rows and l columns. The element c_{ij} in C is defined by

$$\sum_{k=1}^{n} a_{ik}b_{kj} = c_{ij} \qquad i = 1, 2, \ldots, m \quad j = 1, 2, \ldots, l \tag{4.42}$$

To illustrate the numerical computations in finding the product of two matrices consider the following example.

Example 4.2

Find the product matrix $C = AB$ if

$$A = \begin{bmatrix} 1 & -1 \\ 0 & 2 \\ 3 & 1 \end{bmatrix} \qquad B = \begin{bmatrix} 2 & 3 \\ 1 & 0 \end{bmatrix}$$

In accordance with Eq. (4.42) the elements of C are given as

$$c_{11} = a_{11}b_{11} + a_{12}b_{21} = 1$$

$$c_{12} = a_{11}b_{12} + a_{12}b_{22} = 3$$

$$c_{21} = a_{21}b_{11} + a_{22}b_{21} = 2$$

$$c_{22} = a_{21}b_{12} + a_{22}b_{22} = 0$$

$$c_{31} = a_{31}b_{11} + a_{32}b_{21} = 7$$

$$c_{32} = a_{31}b_{12} + a_{32}b_{22} = 9$$

Thus the matrix C is

$$C = \begin{bmatrix} 1 & 3 \\ 2 & 0 \\ 7 & 9 \end{bmatrix}$$

The product of two matrices is not defined when the number of columns in the first is not equal to the number of rows in the second.

On the basis of the definitions above an algebra of matrices can be formulated and is summarized below. Lowercase letters denote scalar quantities and uppercase letters denote matrices. Each entry in this list is readily proved through the use of the foregoing definitions.

a. A matrix times a zero matrix is a zero matrix.

b. The product of the matrix A and the unit matrix I is the matrix A:

$$AI = IA = A$$

c. Addition of matrices is associative:

$$(A + B) + C = A + (B + C) = A + B + C$$

d. Addition of matrices is commutative:

$$A + B = B + A$$

e. Adding a zero matrix does not change a matrix:

$$A + 0 = A$$

f. Multiplication by a scalar is distributive and associative:

$$(a + b)A = aA + bA$$
$$a(A + B) = aA + aB$$
$$aAB = AaB = ABa$$

g. Multiplication of matrices is distributive:

$$(A + B)C = AC + BC$$
$$A(B + C) = AB + AC$$

h. Matrix multiplication is associative:

$$(AB)C = A(BC) = ABC$$

i. Multiplication of matrices is *not* commutative:

$$AB \neq BA$$

Each of the algebraic properties listed above is easily verified by performing the operations according to definitions 1, 2, and 3 above.

The determinant of a square matrix A is that determinant formed by the elements of A. This determinant, written det A or $|A|$, is called the characteristic determinant of A:

$$\det A = |A| = \begin{vmatrix} a_{11} & a_{12} & \cdots & a_{1n} \\ a_{21} & a_{22} & \cdots & a_{2n} \\ \cdots & \cdots & \cdots & \cdots \\ a_{n1} & a_{n2} & \cdots & a_{nn} \end{vmatrix} \tag{4.43}$$

If the characteristic determinant of A is equal to zero, the matrix A is called a singular matrix. Conversely, if det $A \neq 0$, then A is a nonsingular matrix.

The determinant of the transpose of a matrix is equal to the determinant of the matrix:

$$\det A^t = \det A$$

The product of the determinants of two n by n matrices A and B is the determinant of the product AB:

$$\det A \cdot \det B = \det AB$$

The proof of the last two statements follows from the properties listed above and the properties of determinants.

With each nonsingular matrix A there exists a matrix, called the *inverse matrix* and written A^{-1}, which is defined by

$$A^{-1}A = AA^{-1} = I \tag{4.44}$$

The inverse of A, A^{-1}, is defined as having elements α_{ij} as follows:

$$A^{-1} = [\alpha_{ij}] = \begin{bmatrix} \alpha_{11} & \alpha_{12} & \cdots & \alpha_{1n} \\ \alpha_{21} & \alpha_{22} & \cdots & \alpha_{2n} \\ \cdots & \cdots & \cdots & \cdots \\ \alpha_{n1} & \alpha_{n2} & \cdots & \alpha_{nn} \end{bmatrix} \tag{4.45}$$

where $\alpha_{ij} = A_{ji}/\det A$. The term A_{ji} is the cofactor of the element a_{ji} in A. The product of A and A^{-1} in either order is the unit matrix I. An example illustrates finding the inverse of A.

Example 4.3

Find the inverse of the coefficient matrix in Eqs. (4.6).

$$A = [a_{ij}] = \begin{bmatrix} 2 & 4 & 2 \\ 2 & -1 & -2 \\ 4 & 1 & -2 \end{bmatrix}$$

$$A^{-1} = [\alpha_{ij}]$$

$$\alpha_{11} = \frac{\begin{vmatrix} -1 & -2 \\ 1 & -2 \end{vmatrix}}{\det A}(-1)^2 = 4/4 = 1$$

$$\alpha_{12} = \frac{\begin{vmatrix} 4 & 2 \\ 1 & -2 \end{vmatrix}}{4}(-1)^3 = 10/4$$

$$\alpha_{13} = \frac{\begin{vmatrix} 4 & 2 \\ -1 & -2 \end{vmatrix}}{4}(-1)^4 = -6/4$$

$$\alpha_{21} = \frac{\begin{vmatrix} 2 & -2 \\ 4 & -2 \end{vmatrix}}{4}(-1)^3 = -4/4$$

$$\alpha_{22} = \frac{\begin{vmatrix} 2 & 2 \\ 4 & -2 \end{vmatrix}}{4}(-1)^4 = -12/4$$

$$\alpha_{23} = \frac{\begin{vmatrix} 2 & 2 \\ 2 & -2 \end{vmatrix}}{4}(-1)^5 = 8/4$$

$$\alpha_{31} = \frac{\begin{vmatrix} 2 & -1 \\ 4 & 1 \end{vmatrix}}{4}(-1)^4 = 6/4$$

$$\alpha_{32} = \frac{\begin{vmatrix} 2 & 4 \\ 4 & 1 \end{vmatrix}}{4}(-1)^5 = 14/4$$

$$\alpha_{33} = \frac{\begin{vmatrix} 2 & 4 \\ 2 & -1 \end{vmatrix}}{4}(-1)^6 = -10/4$$

Thus the inverse of A is

$$A^{-1} = \frac{1}{4}\begin{bmatrix} 4 & 10 & -6 \\ -4 & -12 & 8 \\ 6 & 14 & -10 \end{bmatrix}$$

4.14 *Matrix Equations*

The equation

$$AX = B \tag{4.46}$$

where A, X, and B are matrices that are properly defined, is a matrix equation. From the rules of matrix multiplication given above the number of columns in the A matrix must equal the number of rows in the X matrix. The B matrix will contain the same number of rows as the A matrix and the same number of columns as the X matrix. The matrix equation written in an expanded form is

$$\begin{bmatrix} a_{11} & a_{12} & \cdots & a_{1n} \\ a_{21} & a_{22} & \cdots & a_{2n} \\ \cdots & \cdots & \cdots & \cdots \\ a_{m1} & a_{m2} & \cdots & a_{mn} \end{bmatrix} \begin{bmatrix} x_{11} & \cdots & x_{1l} \\ x_{21} & \cdots & x_{2l} \\ \cdots & \cdots & \cdots \\ x_{n1} & \cdots & x_{nl} \end{bmatrix} = \begin{bmatrix} b_{11} & b_{12} & \cdots & b_{1l} \\ b_{21} & b_{22} & \cdots & b_{2l} \\ \cdots & \cdots & \cdots & \cdots \\ b_{m1} & b_{m2} & \cdots & b_{ml} \end{bmatrix} \tag{4.47}$$

The elements in the B matrix satisfy

$$\sum_{p=1}^{n} a_{ip}x_{pk} = b_{ik} \tag{4.48}$$

In Eq. (4.47) A is an m by n matrix, X is an n by l matrix, and B is an m by l matrix. If the elements of A and B are assumed known, and if l is equal to 1, the matrix equation simply represents a set of m simultaneous linear equations in n unknown values of x which are the elements of an n-dimensional vector X. This is clear from the definition of the elements of the B matrix in Eq. (4.48). For if l is equal to 1, the subscript k can only be equal to 1 and can be dropped. The result is

$$\sum_{j=1}^{n} a_{ij}x_j = b_i \tag{4.49}$$

This equation holds for $i = 1, 2, \ldots, m$, and is therefore a set of m equations in n unknowns. If m is equal to n, the A matrix is a square matrix, and there will be n simultaneous equations in the n elements of X. Under these conditions the matrix equation $AX = B$ represents a set of n equations in n unknowns.

For the case where l is greater than 1 and A is an n by n matrix, the equation for each element in the kth column of B is

$$\sum_{j=1}^{n} a_{ij}x_{jk} = b_{ik} \qquad \text{for } i = 1, 2, \ldots, n \tag{4.50}$$

This is still a set of n equations in the n unknowns $x_{1k}, x_{2k}, \ldots, x_{nk}$, which are the elements in the kth column of X. Thus, Eqs. (4.50) represent l sets of simultaneous equations in the unknown elements of the matrix X. Each set of equations can be solved by methods already presented.

The characteristic determinant for the set of equations is the determinant of the matrix A, det A. If A is singular, i.e., det $A = 0$, it is impossible to find a unique solution for the matrix X. On the other hand, if A is nonsingular, i.e., det $A \neq 0$, a unique solution does exist. In this case, the set of equations in (4.50) with $k = 1$ can be solved for the first column of the X matrix by methods already discussed. Then the set for $k = 2$ can be solved for the second column of X, etc. In this manner, column by column, the complete solution for the X matrix can be obtained.

A considerable saving in computational effort can be made, however, by performing the solution for all elements of the X matrix at the same time. This can be done by first adjoining the B matrix to the A matrix

$$\begin{bmatrix} a_{11} & a_{12} & \cdots & a_{1n} & b_{11} & b_{12} & \cdots & b_{1m} \\ a_{21} & a_{22} & \cdots & a_{2n} & b_{21} & b_{22} & \cdots & b_{2m} \\ \cdot & \cdot & \cdot & \cdot & \cdot & \cdot & \cdot & \cdot \\ a_{n1} & a_{n2} & \cdots & a_{nn} & b_{n1} & b_{n2} & \cdots & b_{nm} \end{bmatrix} \tag{4.51}$$

and then using elimination on the entire matrix. If the Gauss-Jordan elimination method is used, the matrix equation will be in the form $IX = B'$ after all computations have been completed. The solution matrix X will be in the B' matrix in the position originally occupied by the B matrix.

$$\begin{bmatrix} 1 & 0 & \cdots & 0 & x_{11} & x_{12} & \cdots & x_{1m} \\ 0 & 1 & \cdots & 0 & x_{21} & x_{22} & \cdots & x_{2m} \\ \cdot & \cdot & \cdot & \cdot & \cdot & \cdot & \cdot & \cdot \\ 0 & 0 & \cdots & 1 & x_{n1} & x_{n2} & \cdots & x_{nm} \end{bmatrix} \tag{4.52}$$

A flow chart using the Gauss-Jordan elimination method to solve a matrix equation is shown in Fig. 4.4. Reference should be made to Sec. 4.9 for an explanation of the Gauss-Jordan method. Note in statements 4 and 9 that the iteration begins with the column index set to $k + 1$. This means that the specific computations to place unity on the diagonal and to place zeros in the column positions are not done, which yields a small saving in computer time.

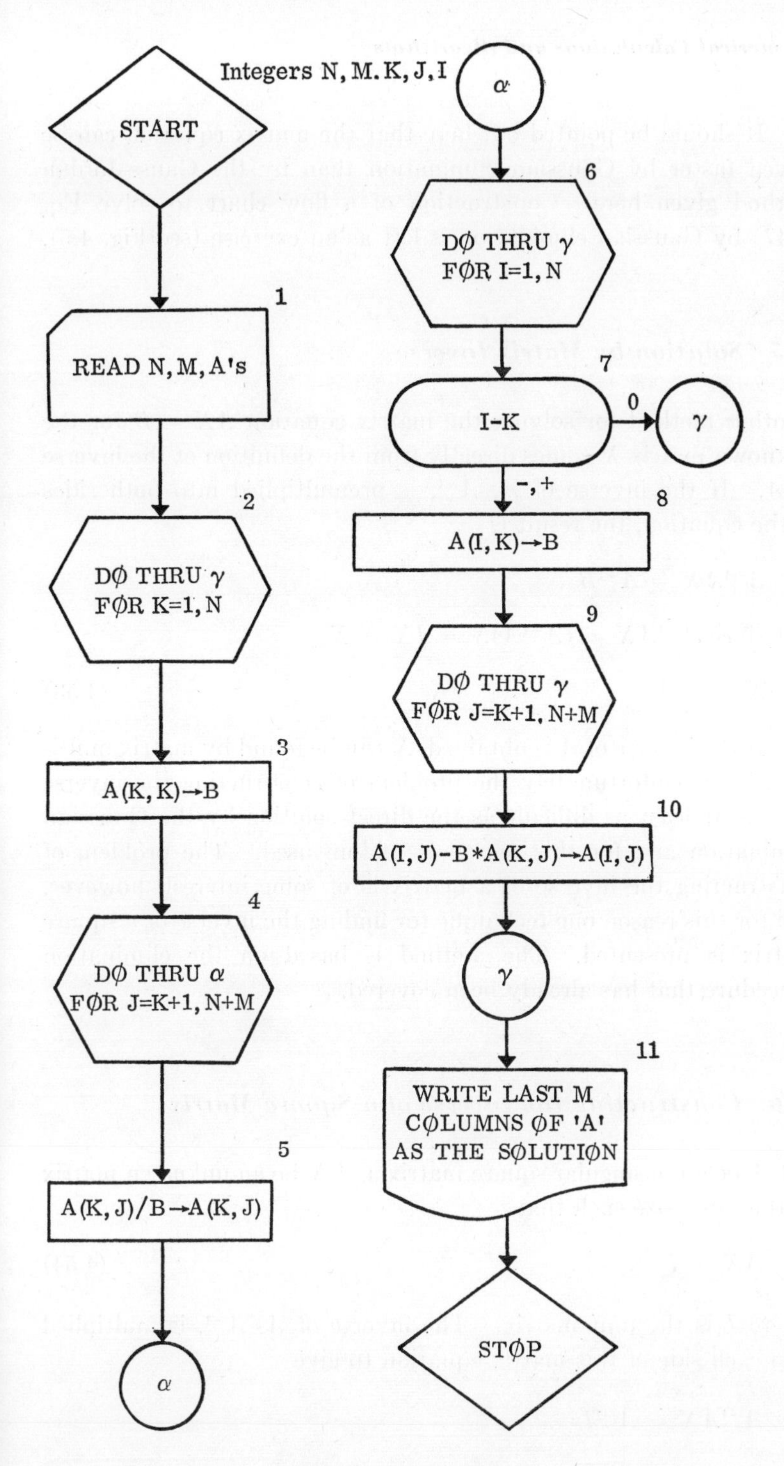

Figure 4.4 Gauss-Jordan elimination to solve a matrix equation.

It should be pointed out here that the matrix equation can be solved faster by Gaussian elimination than by the Gauss-Jordan method given here. Construction of a flow chart to solve Eq. (4.47) by Gaussian elimination is left as an exercise (see Fig. 4.5).

4.15 Solution by Matrix Inverse

Another method for solving the matrix equation $AX = B$ for the unknown matrix X comes directly from the definition of the inverse of A. If the inverse of A, A^{-1}, is premultiplied into both sides of the equation, the result is

$$A^{-1}AX = A^{-1}B$$

But since $A^{-1}AX = (A^{-1}A)X = IX = X$,

$$X = A^{-1}B \tag{4.53}$$

If the inverse of A is obtained, X can be found by matrix multiplication. Unfortunately the problem of constructing the inverse of A is at least as difficult as the direct solution by the Gaussian elimination and for this reason is seldom used. The problem of constructing the inverse of a matrix is of some interest, however, and for this reason one technique for finding the inverse of a square matrix is presented. The method is based on the elimination procedure that has already been covered.

4.16 Constructing the Inverse of a Square Matrix

Let A be a nonsingular square matrix and X be an unknown matrix of the same size such that

$$AX = I \tag{4.54}$$

where I is the unit matrix. The inverse of A, A^{-1}, is multiplied into each side of this matrix equation to give

$$A^{-1}AX = A^{-1}I$$

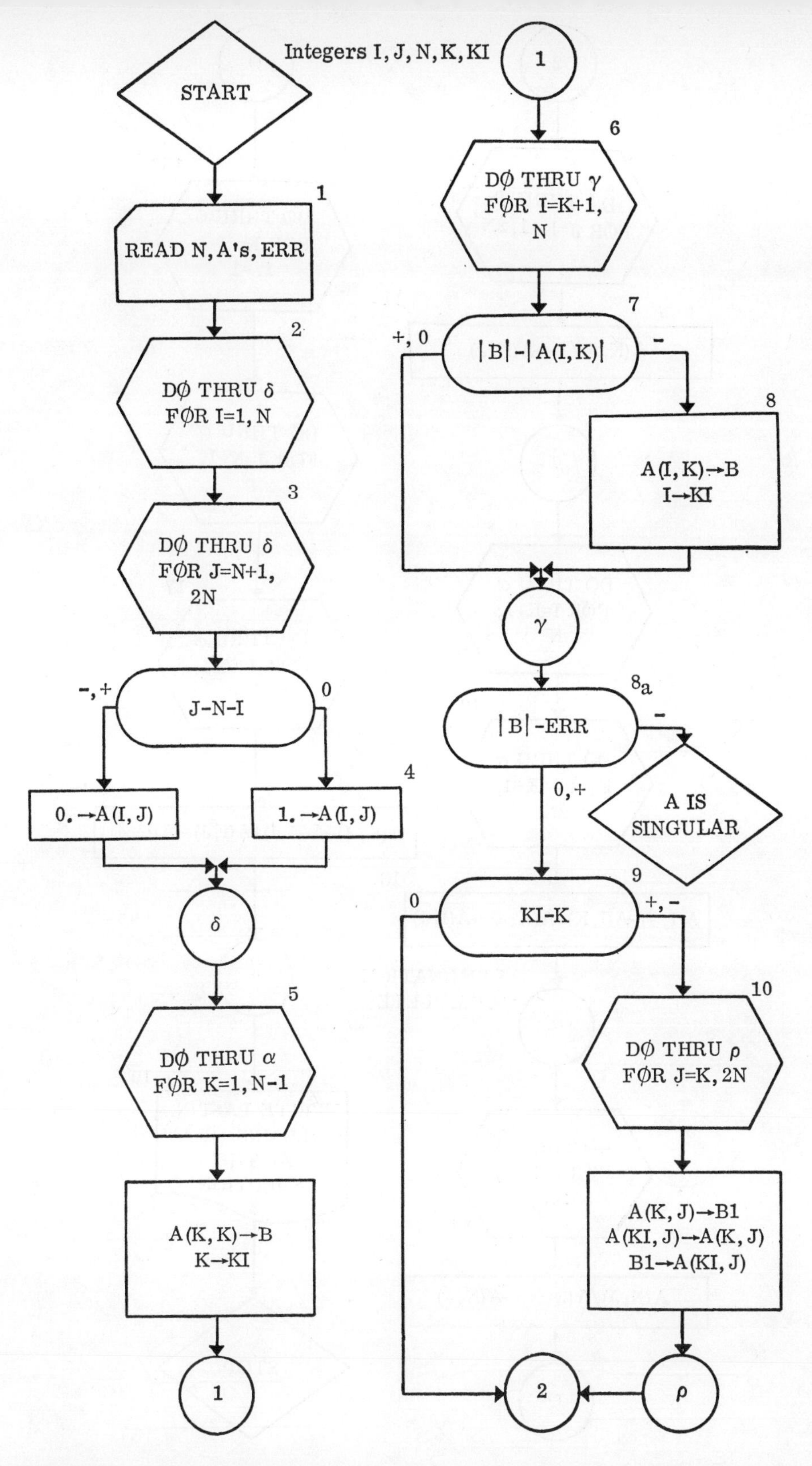

Figure 4.5 Invert the square matrix *A*.

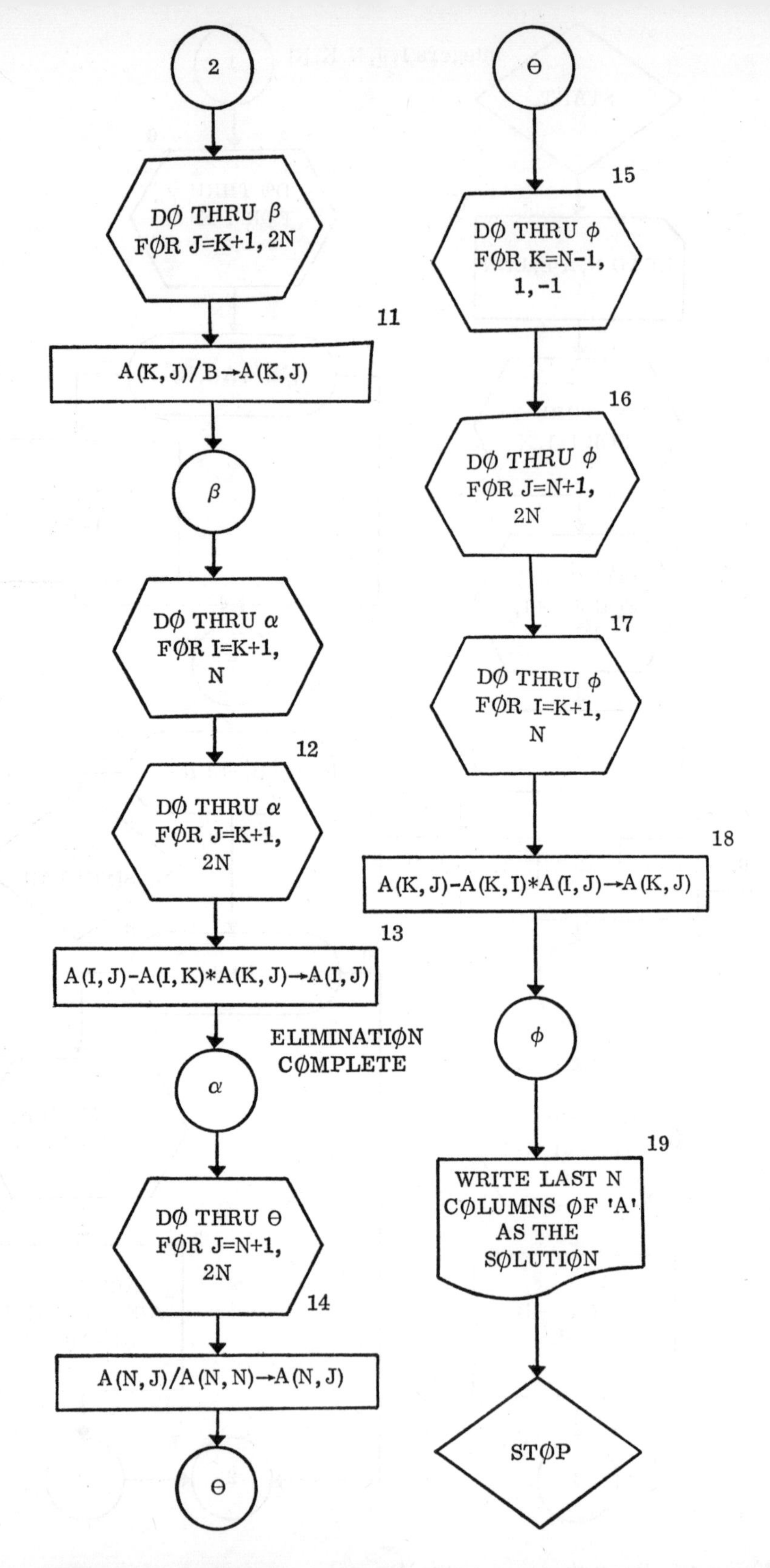

Figure 4.5 *(Continued)*

Since $A^{-1}AX = (A^{-1}A)X = IX = X$, and also $A^{-1}I = A^{-1}$, then

$$X = A^{-1} \tag{4.55}$$

The solution for X in the equation $AX = I$ is the inverse of A.

Any method used to solve the equation $AX = B$ can be used to solve $AX = I$. The only change in the routine is that the unit matrix I, which is the same size as A, replaces the matrix B in the computations.

A flow chart for finding the inverse of A when A is nonsingular is given in Fig. 4.5. The method used in the flow chart is Gaussian elimination with pivoting. A short explanation of the steps in the flow chart is given.

2–4. The unit matrix is adjoined to the square matrix A and is placed in locations A(I, N + J), I, J = 1, 2, . . . , n.

5–10. Statement 5 sets control to eliminate the kth unknown from all equations (or rows) below the kth row. Statements 6 through 10 perform a pivoting operation which finds the largest element in column k for rows k to n. The row containing the largest element is then interchanged with row k in the iteration controlled by 10. The logic in 8a identifies a singular matrix when ERR is appropriately defined. Statement 9 bypasses the row interchange when row k already contains the largest available value.

11–13. These statements perform the elimination and leave the A matrix in the form shown in Eq. (4.18). Note that the elimination steps are carried to $2n$ on the column index.

14–19. These steps include the back substitution necessary to construct the solution matrix. The solution matrix X [see Eq. (4.55)] is placed into locations A(I, J + N), I, J = 1, 2, . . . , N. This gives some economy in storage space and is important if the size of memory is a limitation.

4.17 Matrix Eigenvalue Problems

The matrix equation

$$AX = Y$$

where A is an n by n matrix and X and Y are n-dimensional vectors, is a linear transformation of the vector X into the vector Y. An important computational problem is to find a vector X which is transformed by the matrix A into a vector that is collinear with X. This problem may be stated as

$$AX = Y = \lambda X \tag{4.56}$$

The vectors Y and X are in the same direction, and thus are said to be collinear, but they differ in magnitude by the scalar multiplier λ. The vector X that will satisfy Eq. (4.56) is called an *eigenvector* of the matrix A, and the corresponding value for λ is called an *eigenvalue* of A. The eigenvalue problem has wide application in the sciences and engineering and is discussed in some detail. An example is solved for illustration.

Example 4.4

A simple example of an eigenvalue problem is that of finding the principal moments of inertia of a plane figure. Let the inertia components referred to an xy axis be given as $H_x = 5$, $H_y = 5$, and $H_{xy} = H_{yx} = 3$. The principal moments of inertia and the directions called *principal directions,* about which these moments occur, are defined by the matrix equation

$$HX = \lambda X$$

where H is the matrix formed by the inertia components (see standard texts on statics).

$$H = \begin{vmatrix} H_x & -H_{xy} \\ -H_{xy} & H_y \end{vmatrix} = \begin{vmatrix} 5 & -3 \\ -3 & 5 \end{vmatrix}$$

The eigenvalue λ and the eigenvector X give a principal moment of inertia and its principal direction, respectively. The matrix equation is written in the following expanded form, where the vector components x_1 and x_2 are along x and y, respectively:

$$\begin{vmatrix} 5 & -3 \\ -3 & 5 \end{vmatrix} \begin{vmatrix} x_1 \\ x_2 \end{vmatrix} = \lambda \begin{vmatrix} x_1 \\ x_2 \end{vmatrix} \tag{4.57}$$

This is equivalent to the scalar equations

$$\begin{aligned} 5x_1 - 3x_2 &= \lambda x_1 \\ -3x_1 + 5x_2 &= \lambda x_2 \end{aligned} \tag{4.58}$$

A nonzero solution of the homogeneous equations (4.58) exists if the characteristic determinant is equal to zero.

$$\det (H - \lambda I) = \begin{vmatrix} 5 - \lambda & -3 \\ -3 & 5 - \lambda \end{vmatrix} = 0 \tag{4.58a}$$

Values for λ which satisfy (4.58*a*) are $\lambda = 2$ and $\lambda = 8$. They are the principal moments of inertia. The corresponding directions are found by substituting each λ into Eq. (4.58) and solving for x_1 and x_2. For $\lambda = \lambda_1 = 8$, Eqs. (4.58) give

$$-3x_1 - 3x_2 = 0 \qquad \text{or} \qquad x_1 = -x_2$$

This defines a direction $-45°$ from the x axis, i.e., the direction of the axis having the largest principal moment. Note that unique values for x_1 and x_2 are not defined. Additional conditions are necessary to make the vector unique, and they are usually imposed on the length and sense of the vector or one of its components. For example, if x_1 is chosen 1 unit long and in the positive sense, the vector is defined uniquely.

In a similar way it is found that the eigenvector for $\lambda = \lambda_2 = 2$ is $x_1 = 1$, $x_2 = 1$. This defines a direction of 45° from the x axis.

Equation (4.56) is a linear homogeneous equation in X which can be written

$$AX - \lambda X = (A - \lambda I)X = 0 \tag{4.59}$$

one solution for which is the trivial solution, $X = 0$. A nontrivial solution exists only when the characteristic determinant is equal to zero.

$$\det(A - \lambda I) = \begin{vmatrix} a_{11} - \lambda & a_{12} & \cdots & a_{1n} \\ a_{21} & a_{22} - \lambda & \cdots & a_{2n} \\ \cdots & \cdots & \cdots & \cdots \\ a_{n1} & a_{n2} & \cdots & a_{nn} - \lambda \end{vmatrix} = 0 \tag{4.60}$$

This determinant represents an nth-degree polynomial in λ. In general, there are n values of λ which satisfy Eq. (4.60). These n values, noted λ_i, $i = 1, \ldots, n$, are the n eigenvalues for the matrix A. For each eigenvalue λ_i there is a corresponding nonzero eigenvector X_i that is found from

$$AX_i = \lambda_i X_i \tag{4.61}$$

The set of equations represented by Eq. (4.61) is linearly dependent and homogeneous, so that the solution for the eigenvector cannot be unique. However, if λ_i is not a multiple root of Eq. (4.60), the components of the eigenvector X_i can be determined to a constant multiplier. If the nth component of the vector X_i is set equal to the constant $x_{n,i}$, each of the other components will be uniquely defined in terms of this constant:

$$\begin{aligned} x_{1,i} &= \alpha_{1,i} \quad x_{n,i} \\ &\cdots\cdots \\ x_{n-1,i} &= \alpha_{n-1,i} \quad x_{n,i} \\ x_{n,i} &= x_{n,i} \end{aligned} \tag{4.62}$$

The value of $x_{n,i}$ is arbitrary and is often set equal to 1, in which case the vector X_i is given by

$$X_i = \begin{bmatrix} \alpha_{1,i} \\ \alpha_{2,i} \\ \cdot \\ \cdot \\ \cdot \\ \alpha_{n-1,i} \\ 1 \end{bmatrix} \tag{4.63}$$

4.18 Properties of Eigenvalues and Eigenvectors

Many interesting and useful properties are exhibited by the eigenvectors and eigenvalues of a matrix. Only those that are pertinent to the method of solution proposed here will be discussed; reference is made to the bibliography for further reading.

1. Each eigenvector of a matrix A is unique except for a constant multiplier. If the vector X_i satisfies Eq. (4.59), the vector cX_i, where c is any scalar, also satisfies this equation. If c is positive, the sense of cX_i is the same as that of X_i; whereas if c is negative, the sense of cX_i is opposite that of X_i. Thus, the eigenvector X_i is unique if its magnitude and sense are prescribed.

2. The inverse of an eigenvalue of the matrix A is equal to an eigenvalue of the inverse of A. This is illustrated as follows. Let λ_i be an eigenvalue of the matrix A:

$$AX_i = \lambda_i X_i \tag{4.61}$$

Multiply both sides of this equation by the inverse of A (for A nonsingular) and divide by λ_i:

$$\frac{1}{\lambda_i} X_i = A^{-1}X_i \tag{4.64}$$

From this equation the eigenvalues of A^{-1} are $1/\lambda_i$, and the statement is verified. The corresponding eigenvectors will be the same since X_i is not altered.

3. If two n by n matrices A and B are related through the nonsingular matrix H,

$$B = HAH^{-1} \tag{4.65}$$

then the eigenvalues of the matrix A are equal to the eigenvalues of the matrix B. This transformation of the matrix A into the matrix B is called a *similarity transformation*, and the statement is proved as follows.

Multiply the matrix H into both sides of Eq. (4.61) and replace X_i in the left-hand side by the equivalent expression, $H^{-1}HX_i$, giving

$$HAH^{-1}HX_i = \lambda_i HX_i \tag{4.66}$$

Substituting from Eq. (4.65) and setting $Y_i = HX_i$ gives

$$BY_i = \lambda_i Y_i \tag{4.67}$$

The eigenvalue λ_i is not altered in the operations above. Therefore an eigenvalue for Eq. (4.67) is equal to an eigenvalue for Eq. (4.61). Since i is arbitrary, each eigenvalue of matrix A is an eigenvalue of matrix B. The eigenvectors of B are equal to the eigenvectors of A multiplied by the matrix H:

$$Y_i = HX_i \tag{4.68}$$

Properties 1 through 3 are helpful in developing an iterative method for finding the eigenvalues and the eigenvectors of a matrix.

4.19 The Power Method

The *power method* is a numerical method for finding an approximation to the largest eigenvalue and the corresponding eigenvector of Eq. (4.61). It is based on selecting a set of starting values for the unknown components of the eigenvector, and by a systematic iteration using Eq. (4.61) the approximation to the eigenvector is improved. When a suitable approximation to the eigenvector is found, an approximate value for the eigenvalue can be obtained.

Let A be an n by n matrix with real eigenvalues λ_i, $i = 1, 2, \ldots, n$, which are arranged so that $|\lambda_i| > |\lambda_{i+1}|$ for each i. A vector $Y^{(0)}$ with elements $y_i^{(0)}$ is selected as the initial approximation to the eigenvector X_1 and is substituted into the left-hand side of Eq. (4.61). This gives the first approximation to the eigenvector X_1 and is labeled $Y^{(1)}$ in

$$\lambda^{(1)} Y^{(1)} = A Y^{(0)} \tag{4.69}$$

The scalar $\lambda^{(1)}$ is included in the equation to control the magnitude of Y. The kth approximation to the eigenvector is obtained from

$$\lambda^{(k)} Y^{(k)} = A Y^{(k-1)} \tag{4.70}$$

The value for $\lambda^{(k)}$ is often chosen as the largest element of $A Y^{(k-1)}$ and is an approximation to the eigenvalue λ_1. If the kth iteration gives a satisfactory approximation to the eigenvector X_1, then $Y^{(k)}$ will satisfy

$$A Y^{(k)} = \lambda^{(k+1)} Y^{(k+1)} = \lambda_1 Y^{(k)} \tag{4.71}$$

It can be shown that this iteration will converge to the eigenvector corresponding to the largest eigenvalue λ_1. See Ref. 10 for a proof of the method.

The steps in the iteration procedure for finding the largest eigenvalue and the corresponding eigenvector are summarized.

1. Choose values for the elements of the initial vector $Y^{(0)}$.

2. Compute the vector $Y^{(k)}$ from the equation

$$\lambda^{(k)} Y^{(k)} = A Y^{(k-1)}$$

first for $k = 1$, then for $k = 2$, and so forth. The value for $\lambda^{(k)}$ is a constant that is chosen to make one of the elements in $Y^{(k)}$ equal to 1. The same element should be set to unity in each iteration.

3. Test the value for $\lambda^{(k)}$ against the value for $\lambda^{(k-1)}$ and/or the value of each element of the vector $Y^{(k)}$ with the corresponding element of $Y^{(k-1)}$. If the change in any of the numbers is greater than a given error criterion, increment k by 1 and return to step 2. If the error criterion is satisfied, the largest eigenvalue is $\lambda^{(k)}$, and the vector $Y^{(k)}$ is the corresponding eigenvector.

Example 4.5

The 3 by 3 matrix A is solved by the power method for the largest eigenvalue and the corresponding eigenvector.

$$A = \begin{bmatrix} 10 & -3 & 2 \\ 1 & 5 & 0 \\ 3 & 1 & 1 \end{bmatrix}$$

The solution is begun by assuming starting components for the eigenvector. Let the vector be (1,1,1). The first step in the solution is to multiply this given vector by the matrix A.

$$\begin{bmatrix} 10 & -3 & 2 \\ 1 & 5 & 0 \\ 3 & 1 & 1 \end{bmatrix} \begin{bmatrix} 1 \\ 1 \\ 1 \end{bmatrix} = \begin{bmatrix} 9 \\ 6 \\ 5 \end{bmatrix} = 9 \begin{bmatrix} 1 \\ 0.667 \\ 0.555 \end{bmatrix}$$

The constant factor 9 is an approximate value for the eigenvalue, and the vector components make the first approximation to the corresponding eigenvector. They are used for the second iteration.

$$\begin{bmatrix} 10 & -3 & 2 \\ 1 & 5 & 0 \\ 3 & 1 & 1 \end{bmatrix} \begin{bmatrix} 1 \\ 0.667 \\ 0.555 \end{bmatrix} = \begin{bmatrix} 9.108 \\ 4.335 \\ 4.222 \end{bmatrix} = 9.108 \begin{bmatrix} 1 \\ 0.475 \\ 0.464 \end{bmatrix}$$

The new approximation for the eigenvalue is 9.108 and the eigenvector is (1.00, 0.475, 0.465). This result is not correct because it does not agree with the previous iteration. The iterations are continued until two consecutive values for the eigenvalue and/or eigenvector are equal or until the difference is not greater than a tolerable error. After eight repetitions with slide-rule accuracy, the following results are obtained:

$$\text{Eigenvalue} = 10.10 \qquad \text{eigenvector} = (1.00, 0.20, 0.35)$$

A flow chart for finding the largest eigenvalue and the corresponding eigenvector of a matrix is given in Fig. 4.6. In this flow chart the largest component is factored from the eigenvector in each iteration.

4.20 *Finding Additional Eigenvalues*

The method presented above will find only the largest eigenvalue and the corresponding eigenvector of a matrix. In order to find another eigenvalue of the same matrix by the power method it is necessary to remove (or factor) the largest value from the matrix.

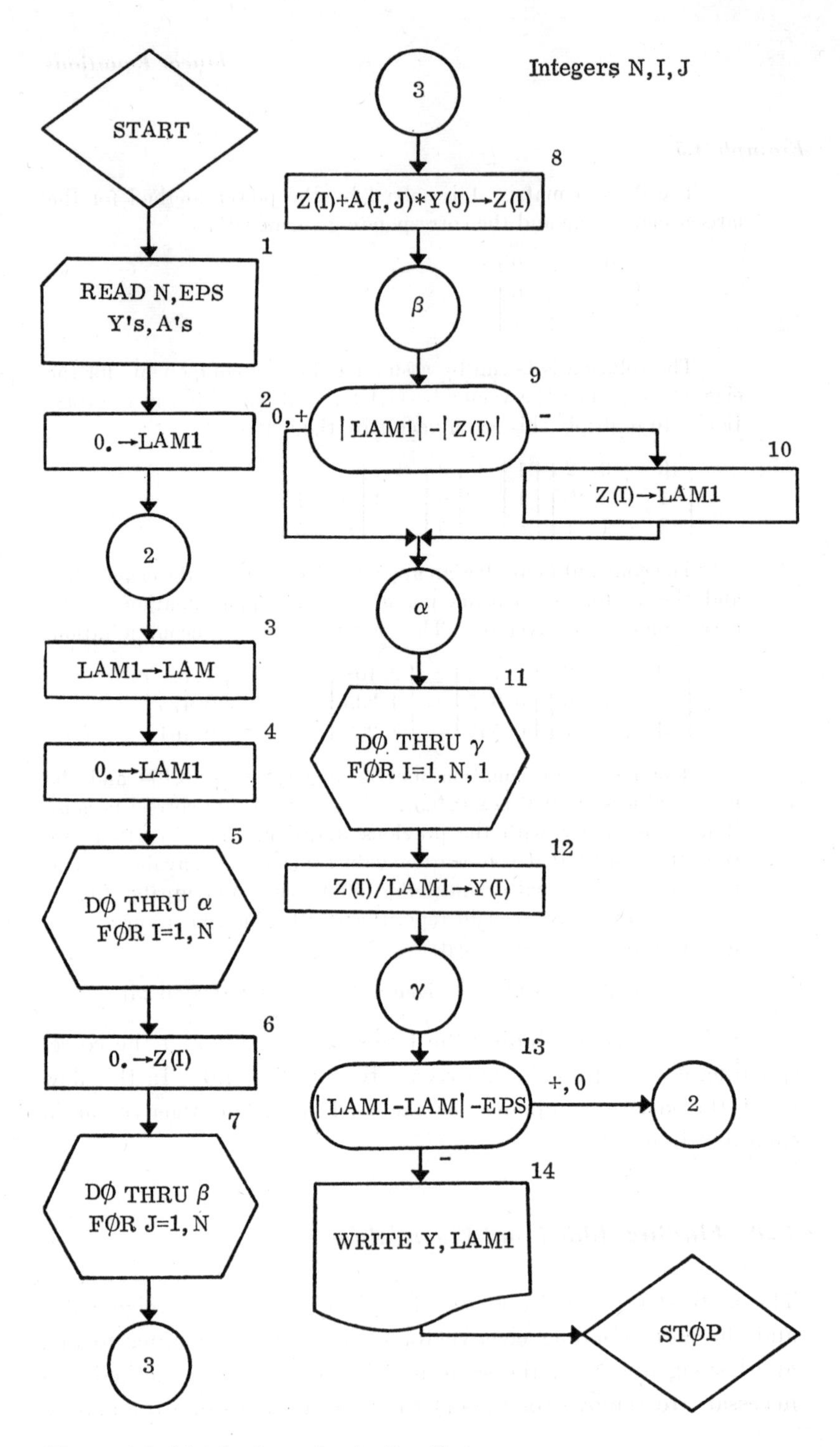

Figure 4.6 Matrix eigenvalue by iteration.

The problem is somewhat similar to that of finding the roots of a polynomial. If one root of an nth-degree polynomial is found, it can be factored from the polynomial to give an $(n - 1)$st-degree polynomial, which will have the same roots as the original except for the one root that has been factored out. A corresponding operation is performed with matrices.

Assume that the largest eigenvalue λ_1 and the corresponding eigenvector X_1 of the n by n matrix A have been found. The matrix is then reduced to an $n - 1$ by $n - 1$ matrix which has the same eigenvalues as A except for λ_1. The largest eigenvalue of the reduced matrix which is the second largest eigenvalue of A can then be obtained by applying the power method to the reduced matrix.

The process whereby the reduced matrix is obtained is called a *similarity transformation* (see property 3 in Sec. 4.18). The elements of a matrix H that will accomplish this similarity transformation are given in terms of the elements of the eigenvector X_1. The matrix H is defined in

$$H = \begin{bmatrix} \dfrac{x_{11}}{x_{11}} & 0 & \cdots & 0 \\ \dfrac{x_{21}}{x_{11}} & 1 & \cdots & 0 \\ \cdots & \cdots & \cdots & \cdots \\ \dfrac{x_{n1}}{x_{11}} & 0 & \cdots & 1 \end{bmatrix} \tag{4.72}$$

where

$$\begin{bmatrix} x_{11} \\ x_{21} \\ \cdot \\ \cdot \\ \cdot \\ x_{n1} \end{bmatrix} = X_1$$

If the magnitude of the eigenvector X_1 is chosen so that $x_{11} = 1$, then H is given by

$$H = \begin{bmatrix} 1 & 0 & \cdots & 0 \\ x_{21} & 1 & \cdots & 0 \\ \cdots & \cdots & \cdots & \cdots \\ x_{n1} & 0 & \cdots & 1 \end{bmatrix} \tag{4.73}$$

A similarity transformation is made on the matrix A with the matrix H and its inverse

$$H^{-1}AH = B \tag{4.74}$$

This defines a new matrix, designated B, which has the same eigenvalues as A. Furthermore because of the special choice of H the eigenvalue and eigenvector which have been used to form H are "factored" from the A matrix. An $n - 1$ by $n - 1$ matrix is obtained which has the same eigenvalues as the matrix A except for the one eigenvalue corresponding to the eigenvector used to form H. The eigenvectors of the transformed matrix are linearly related to the eigenvectors of the original A matrix. This is shown as follows.

In the eigenvalue problem

$$AX = \lambda X$$

insert the unit matrix $HH^{-1} = I$ between A and X and premultiply by H^{-1}.

$$(H^{-1}AH)H^{-1}X = \lambda H^{-1}X \tag{4.75}$$

If Y is set equal to $H^{-1}X$, then

$$BY = \lambda Y$$

and

$$X = HY \tag{4.76}$$

The process of reducing the matrix in this way in order to find additional eigenvalues and eigenvectors is called *deflation*. The proof that the choice for H in Eq. (4.72) leads to deflation of A is difficult and is not attempted here. Instead it is shown that the similarity transformation defined in Eqs. (4.73) and (4.74) leads to the desired result.

The inverse of H is readily constructed from the definition of the inverse. Let h_{ij}^{-1} represent the elements of H^{-1}; then

$$h_{ij}^{-1} = \frac{H_{ji}}{\det H} \tag{4.77}$$

where H_{ji} is the cofactor of h_{ji} in the H matrix. Carrying out the operations in (4.77) gives

$$H^{-1} = \begin{bmatrix} 1 & 0 & \cdots & 0 \\ -x_{21} & 1 & \cdots & 0 \\ -x_{31} & 0 & \cdots & 0 \\ \cdots & \cdots & \cdots & \cdots \\ -x_{n1} & 0 & \cdots & 1 \end{bmatrix} \tag{4.78}$$

The product $H^{-1}AH$ is obtained by carrying out the indicated matrix multiplication and is given in

$$B = H^{-1}AH = \begin{bmatrix} \lambda_1 & a_{12} & \cdots & a_{1n} \\ 0 & a_{22} - a_{12}x_{21} & \cdots & a_{2n} - a_{1n}x_{21} \\ \cdots & \cdots & \cdots & \cdots \\ 0 & a_{i2} - a_{12}x_{i1} & \cdots & a_{in} - a_{1n}x_{i1} \\ \cdots & \cdots & \cdots & \cdots \\ 0 & a_{n2} - a_{12}x_{n1} & \cdots & a_{nn} - a_{1n}x_{n1} \end{bmatrix} \tag{4.79}$$

The general form for the elements in the B matrix is

$$b_{ij} = a_{ij} - a_{1j}x_{i1} \qquad \text{for } i, j = 2, 3, \ldots, n \tag{4.80}$$

The first column in the matrix in Eq. (4.79) is deduced from the equality $AX_1 = \lambda_1 X_1$. The term in position (1,1) of the matrix is $\sum_{j=1}^{n} a_{1j}x_{j1} = \lambda_1 x_{11}$, and since x_{11} is equal to 1, the term is equal to λ_1. The term in position $(k,1)$ is equal to

$$-x_{k1}\Sigma a_{1j}x_{j1} + \Sigma a_{kj}x_{j1}$$

and from $\Sigma a_{kj}x_{j1} = \lambda_1 x_{k1}$ the sum is zero:

$$-x_{k1}\lambda_1 + x_{k1}\lambda_1 = 0$$

λ_1 is clearly an eigenvalue of the matrix in Eq. (4.79).

The reduced $n - 1$ by $n - 1$ matrix is obtained from (4.79) by striking out the first row and the first column. The largest eigenvalue of the reduced matrix can now be computed. It is the second largest eigenvalue of A and is designated λ_2.

The eigenvectors of the reduced matrix are not equal to the eigenvectors of A but are related by the H matrix as shown in Eq. (4.76). Thus, if X_i is the eigenvector of A and Y_i is the corresponding eigenvector of the reduced matrix, then

$$X_i = HY_i \tag{4.81}$$

If the eigenvector of B is designated Y_2, the components of X_2 are given in the following expanded form:

$$\begin{bmatrix} x_{12} \\ x_{22} \\ \cdot \\ \cdot \\ \cdot \\ x_{n2} \end{bmatrix} = \begin{bmatrix} 1 & 0 & \cdots & 0 \\ x_{21} & 1 & \cdots & 0 \\ \cdots & \cdots & \cdots & \cdots \\ \cdots & \cdots & \cdots & \cdots \\ \cdots & \cdots & \cdots & \cdots \\ x_{n1} & 0 & \cdots & 1 \end{bmatrix} \begin{bmatrix} y_{12} \\ y_{22} \\ \cdot \\ \cdot \\ \cdot \\ y_{n2} \end{bmatrix}$$

Thus

$$\begin{aligned} x_{12} &= y_{12} \\ x_{22} &= x_{21}y_{12} + y_{22} \\ x_{32} &= x_{31}y_{12} + y_{32} \\ &\cdots\cdots\cdots \\ x_{n2} &= x_{n1}y_{12} + y_{n2} \end{aligned} \tag{4.82}$$

The first component in the Y_2 vector, i.e., y_{12}, is not determined by the iteration process on the deflated $n - 1$ by $n - 1$ matrix obtained from Eq. (4.79). To obtain this component of Y_2, it is noted that the eigenvector and eigenvalue must satisfy the following equation:

$$BY_2 = \lambda_2 Y_2$$

From this equation and the form for B in Eq. (4.79) it is readily verified that the first component for Y_2, viz., y_{12}, must satisfy the scalar equation

$$\lambda_1 y_{12} + \sum_{j=2}^{n} a_{1j}y_{j2} = \lambda_2 y_{12}$$

Thus

$$y_{12} = \frac{1}{\lambda_2 - \lambda_1} \sum_{j=2}^{n} a_{1j}y_{j2} \tag{4.83}$$

Example 4.6

This numerical example illustrates the iterative method for finding eigenvalues and demonstrates the similarity transformation for deflating the matrix. The following 3 by 3 matrix, designated A, is:

$$A = \begin{bmatrix} 10 & 2 & 0 \\ 9 & 6 & 3 \\ -5 & 2 & 4 \end{bmatrix}$$

All the eigenvalues and eigenvectors of this matrix will be found by using the power method and the deflation procedure outlined.

The initial assumption for the first eigenvector of A is (1,1,1). Table 4.2 gives the approximation for Y and λ for eight consecutive iterations, which are identified in the k column. The first number in the k column indicates the vector and the second the number of iterations to obtain that vector. Convergence to six significant digits is obtained after 25 iterations, and the result is shown opposite 1, 25 in the k column.

The matrix H, which is used in the similarity transformation to deflate the A matrix, is constructed from the first eigenvector, tabulated opposite 1, 25 in the k column of Table 4.2.

$$H = \begin{bmatrix} 1.000000 & 0 & 0 \\ 1.249090 & 1 & 0 \\ -0.294395 & 0 & 1 \end{bmatrix}$$

The deflated matrix is obtained by carrying out the operations in Eq. (4.79) to obtain the B matrix.

$$B = \begin{bmatrix} 12.498181 & 2 & 0 \\ 0 & 3.501819 & 3.000000 \\ 0 & 2.588789 & 4.000000 \end{bmatrix}$$

The second-largest eigenvalue of the A matrix is found by applying the power method to the deflated 2 by 2 matrix obtained by striking out the first row and the first column of B. Call this matrix $B1$.

$$B1 = \begin{bmatrix} 3.501819 & 3 \\ 2.588789 & 4 \end{bmatrix}$$

An initial guess of (1,1) is made for an eigenvector of $B1$, and the power method is applied. The results are shown in Table 4.2 opposite 2, 0 to 2, 8 in column k. Convergence is obtained after eight iterations. The third component of the eigenvector is obtained from the

B matrix as follows:

$$\begin{bmatrix} 12.498181 & 2 & 0 \\ 0 & 3.501819 & 3 \\ 0 & 2.588789 & 4 \end{bmatrix} \begin{bmatrix} y_{12} \\ 1.000000 \\ 1.015674 \end{bmatrix} = 6.548840 \begin{bmatrix} y_{12} \\ 1.000000 \\ 1.015674 \end{bmatrix}$$

The unknown component y_{12} must satisfy the equation

$$y_{12} = \frac{2}{6.548840 - 12.498181} = -0.336172$$

The corresponding eigenvector of the original A matrix labeled $X2$ is obtained by using Eq. (4.76). It is tabulated opposite $X2$ in Table 4.2.

Table 4.2

k	y_1	y_2	y_3	λ
1, 0	1	1	1	
1, 1	1.000000	1.500000	0.083333	12.000000
1, 2	1.000000	1.403846	−0.128205	13.000000
1, 3	1.000000	1.330330	−0.211211	12.807692
1, 4	1.000000	1.291271	−0.251502	12.660661
1, 5	1.000000	1.271056	−0.272081	12.582543
1, 6	1.000000	1.260561	−0.282744	12.542113
1, 7	1.000000	1.255090	−0.288301	12.521122
1, 8	1.000000	1.252231	−0.291205	12.510180
....	....	....	....	......
1, 25	1.000000	1.249090	−0.294395	12.498181
2, 0		1.000000	1.000000	
2, 1		1.000000	1.013376	6.501819
2, 2		1.000000	1.015339	6.541948
2, 3		1.000000	1.015625	6.547836
2, 4		1.000000	1.015666	6.548694
2, 5		1.000000	1.015673	6.548818
.....	....	....	....	......
2, 8		1.000000	1.015674	6.548840
Y	−0.336172	1.000000	1.015674	
$X2$	−0.336172	0.580091	1.114641	
3, 0			1.000000	
3, 1			1.000000	0.952979
$Y3^*$		−0.536111	1.000000	
$Y3$	0.092872	−0.536111	0.455478	
$X3$	0.092872	−0.420106	0.428146	0.952979

$$X2 = \begin{bmatrix} x_{12} \\ x_{22} \\ x_{32} \end{bmatrix} = \begin{bmatrix} 1 & 0 & 0 \\ 1.249090 & 1 & 0 \\ -0.294395 & 0 & 1 \end{bmatrix} \begin{bmatrix} -0.336172 \\ 1.000000 \\ 1.015674 \end{bmatrix}$$

$$= \begin{bmatrix} -0.336172 \\ 0.580091 \\ 1.114641 \end{bmatrix}$$

The third and smallest eigenvalue can be found by deflating the 2 by 2 matrix $B1$. This is done in a similar way to the deflation of the larger matrix. A 2 by 2 matrix called $H1$ is constructed using the eigenvector of $B1$ that is tabulated opposite 2, 8 in Table 4.2.

$$H1 = \begin{bmatrix} 1 & 0 \\ 1.015674 & 1 \end{bmatrix}$$

The similarity transformation on $B1$ using the matrix $H1$ and its inverse matrix gives the matrix $B2$.

$$B2 = H1^{-1}B1H1 = \begin{bmatrix} 1 & 0 \\ -1.015674 & 1 \end{bmatrix} \begin{bmatrix} 3.501819 & 3 \\ 2.588789 & 4 \end{bmatrix} \begin{bmatrix} 1 & 0 \\ 1.015674 & 1 \end{bmatrix}$$

$$= \begin{bmatrix} 6.54884 & 3.000000 \\ 0 & 0.952979 \end{bmatrix}$$

The second eigenvalue of A is factored into the (1,1) position in $B2$, and a zero is placed in the first-column position of the second row. The third eigenvalue is directly in the $B2$ matrix and is 0.952979. The eigenvector of $B2$ corresponding to this eigenvalue is labeled $Y3^*$ and is obtained as follows. (The component 1 is obtained from the trivial equation given by striking out the first column and first row of $B2$.)

$$\begin{bmatrix} 6.548840 & 3.000000 \\ 0 & 0.952979 \end{bmatrix} \begin{bmatrix} y_{23}^* \\ 1 \end{bmatrix} = 0.952979 \begin{bmatrix} y_{23}^* \\ 1 \end{bmatrix}$$

$$y_{23}^* = \frac{3}{0.952979 - 6.548840} = -0.536111$$

To find the eigenvector of the original matrix A corresponding to this smallest eigenvalue requires a transformation with the $H1$ matrix and then a transformation with the H matrix. First it is necessary to find the eigenvector of the $B1$ matrix which is equal to $H1\,Y3^*$ and is labeled $Y3$ with elements y_{23} and y_{33}.

$$Y3 = H1\,Y3^* = \begin{bmatrix} 1 & 0 \\ 1.015674 & 1 \end{bmatrix} \begin{bmatrix} -0.536111 \\ 1 \end{bmatrix} = \begin{bmatrix} -0.536111 \\ 0.455478 \end{bmatrix}$$

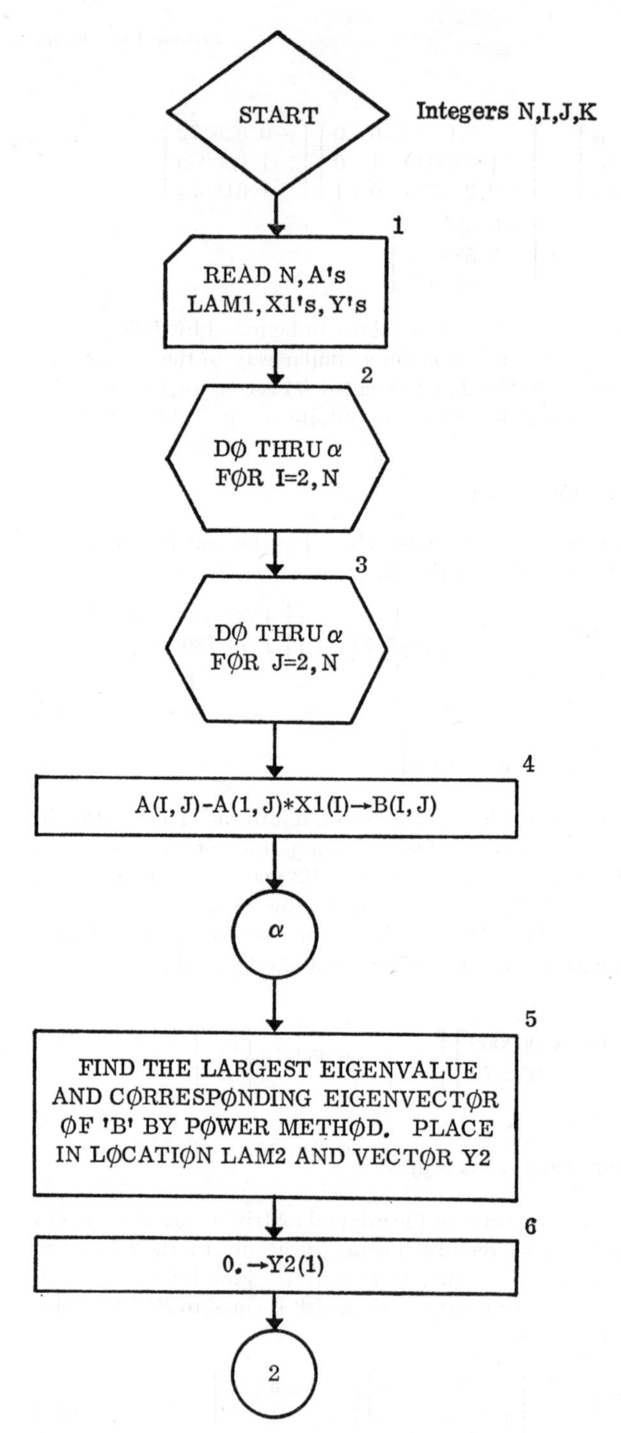

Figure 4.7 Second eigenvalue and eigenvector of the matrix A.

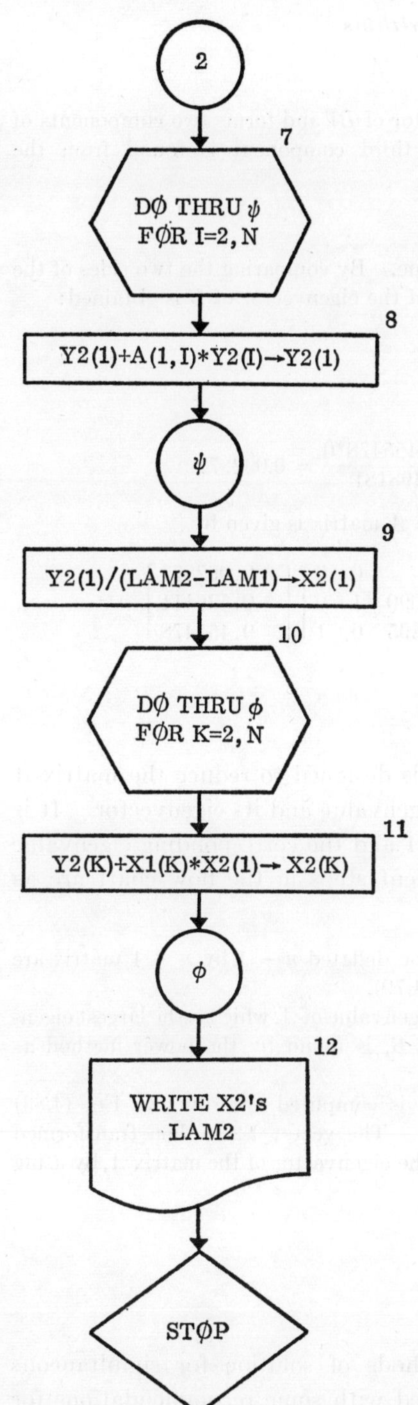

Figure 4.7 *(Continued)*

The vector $Y3$ is an eigenvector of $B1$ and forms two components of an eigenvector of B. The third component is found from the equation

$$B\,Y3 = \lambda_3 Y3$$

where λ_3 is the third eigenvalue. By comparing the two sides of the equation the component y_{13} of the eigenvector of B is obtained:

$$y_{13} = \frac{1}{\lambda_3 - \lambda_1} \sum_{j=2}^{3} b_{ij} y_{j3}$$

$$y_{13} = \frac{-0.536111*2 + 0.455478*0}{0.952979 - 12.498181} = 0.092872$$

Finally the eigenvector of the A matrix is given by

$$X3 = H\,Y3 = \begin{bmatrix} 1 & 0 & 0 \\ 1.249090 & 1 & 0 \\ -0.294395 & 0 & 1 \end{bmatrix} \begin{bmatrix} 0.092872 \\ -0.536111 \\ 0.455478 \end{bmatrix}$$

$$= \begin{bmatrix} 0.092872 \\ -0.420106 \\ 0.428146 \end{bmatrix}$$

The flow chart in Fig. 4.7 is designed to reduce the matrix A and solve for the next-largest eigenvalue and its eigenvector. It is assumed that the eigenvector $X1$ and the corresponding eigenvalue λ_1 are entered as data. Pertinent steps in the flow chart are as follows:

2–4. The elements of the deflated $n - 1$ by $n - 1$ matrix are computed according to Eq. (4.79).

5. The second-largest eigenvalue of A, which is the largest eigenvalue of the deflated matrix B, is found by the power method as shown in Fig. 4.6.

6–11. The value for y_1 is computed according to Eq. (4.83) and stored in location Y2(1). The vector $Y2$ is then transformed into the vector $X2$, which is the eigenvector of the matrix A, by using Eq. (4.76).

4.21 Summary

Elimination and iterative methods of solution for simultaneous algebraic equations are presented with some recommendations for their use. The elimination methods are extended to cover the evaluation of a determinant and the inverse of a matrix. Both the iterative methods and the elimination methods are used to solve linear algebraic matrix equations.

An iterative method, called the power method, is presented for finding the eigenvalues and eigenvectors of a matrix. A deflation scheme is introduced, which permits the computation of all the eigenvalues and eigenvectors by the power method.

Each procedure introduced in the chapter is illustrated by a flow chart with explanation.

Problems

1. Find the solution to the simultaneous equations represented by the following matrices:

(a) $\begin{bmatrix} 1 & 2 & 6 \\ 4 & 3 & 0 \end{bmatrix}$ (b) $\begin{bmatrix} 2 & 3 & 4 & 6 \\ 4 & 7 & 8 & 1 \\ 6 & 9 & 7 & 3 \end{bmatrix}$

(c) $\begin{bmatrix} 1 & 3 & 6 & 2 \\ 2 & 6 & 0 & 1 \\ -5 & 2 & -1 & 3 \end{bmatrix}$ (d) $\begin{bmatrix} 1 & 7 & 2 & 0 & -6 \\ 4 & 6 & 0 & 1 & 8 \\ 3 & 2 & 2 & 1 & -4 \\ 0 & 2 & 9 & 4 & 2 \end{bmatrix}$

2. Write a flow chart to solve p sets of equations similar to those in Prob. 1 by the method of elimination.

3. Do Prob. 2 using Gauss-Seidel iteration.

4. Modify the flow chart in Prob. 2 to test how well each equation in the set is satisfied by the solution that is generated.

5. Solve the following set of equations by Gaussian elimination:

$$\begin{vmatrix} 1 & 1 & 3 & -1 & -2 \\ 2 & 2 & -3 & -1 & 0 \\ 1 & 0 & -1 & -2 & -7 \\ 3 & 0 & 21 & 1 & 7 \end{vmatrix}$$

6. Write a flow chart to solve equations with the properties of Prob. 5 by the method of Gaussian elimination. *Note:* Zero elements on the diagonal must be replaced.

7. Write a flow chart to find the largest candidate for each diagonal position and place it in the diagonal position for the elimination of each unknown. *Hint:* For elimination of the kth unknown do the search in the kth column and also in the kth row. If the largest element is in the row, the columns must be interchanged.

8. Find the value of the characteristic determinant for the system in

(a) Prob. 1a (b) Prob. 1b
(c) Prob. 1c (d) Prob. 1d

9. Evaluate the following determinants by Gaussian elimination:

(a) $\begin{bmatrix} 6 & 3 & 2 \\ 6 & 4 & 3 \\ 20 & 15 & 12 \end{bmatrix}$ (b) $\begin{bmatrix} 1 & 1 & 1 & 1 \\ 1 & 2 & 3 & 4 \\ 1 & 3 & 6 & 10 \\ 1 & 4 & 10 & 20 \end{bmatrix}$

(c) $\begin{bmatrix} 1 & 2 & 1 & 0 \\ 4 & 1 & 3 & 2 \\ 1 & 5 & 0 & 1 \\ 5 & 0 & 5 & 1 \end{bmatrix}$ (d) $\begin{bmatrix} 1 & 1 & 3 & -1 \\ 2 & 2 & -3 & -1 \\ 1 & 0 & -1 & -2 \\ 3 & 0 & 21 & 1 \end{bmatrix}$

10. Write a flow chart that will evaluate n determinants which are m by m, where m is variable.

11. Find the product of the two matrices AB, where

$$A = \begin{bmatrix} 2 & 1 & 0 \\ 1 & 2 & 3 \\ 0 & 2 & 1 \end{bmatrix} \qquad B = \begin{bmatrix} 1 \\ 4 \\ 2 \end{bmatrix}$$

12. Find the product ABC^t, where A and B are given in Prob. 11 and

$$C = \begin{bmatrix} 2 \\ 1 \\ 1 \end{bmatrix}$$

13. A is an n by n matrix, and B and C are vectors defined to make the product ABC^t meaningful. Construct a flow chart to evaluate the product.

14. The matrix A has elements $a_{i,j}$, $i = 1, \ldots, n$, $j = 1, \ldots, m$. The matrix B has element $b_{p,q}$, $p = 1, \ldots, m$, $q = 1, \ldots, l$. Construct a flow chart to find the elements of the product $AB = C$ and store the elements in the matrix C.

15. Find the elements of the product $AA^t = C$.

$$A = \begin{bmatrix} 1 & 2 & 1 & 1 & 2 & 1 \\ 2 & 3 & 5 & 2 & 5 & 3 \\ 3 & 4 & 1 & 1 & 3 & 2 \end{bmatrix}$$

16. Construct a flow chart to find the elements of the product AA^t when A is an n by m matrix.

17. Construct the inverse of the following 3 by 3 matrix by using the definition of the elements of the inverse. Test the result.

$$A = \begin{bmatrix} 2 & 1 & 3 \\ 0 & 6 & 2 \\ 1 & 0 & 1 \end{bmatrix}$$

18. Construct a flow chart for finding the inverse of an n by n matrix with elements $a_{i,j}$ by Gauss-Jordan elimination.

19. Solve the two simultaneous equations by matrix inverse:

$$x_1 + 2x_2 = 5$$

$$2x_1 - x_2 = 0$$

20. Solve Prob. 19 by Gauss-Seidel iteration. Show that the convergence of the iterations depends upon the way in which the equations are arranged.

21. Construct a flow chart to solve Prob. 1 by Gauss-Seidel iteration.

22. Find the largest eigenvalue and the corresponding eigenvector of the following matrices by the power method.

(a) $\begin{bmatrix} 12 & -5 \\ -5 & 6 \end{bmatrix}$ (b) $\begin{bmatrix} 6 & -3 \\ -3 & 2 \end{bmatrix}$

(c) $\begin{bmatrix} 4 & 2 & 1 \\ 3 & 0 & 6 \\ 1 & 3 & 5 \end{bmatrix}$ (d) $\begin{bmatrix} 1{,}000 & -500 & 0 \\ -500 & 1{,}800 & 0 \\ 0 & 0 & 600 \end{bmatrix}$

(e) $\begin{bmatrix} -200 & 100 & -500 \\ 100 & 600 & 0 \\ -500 & 0 & 300 \end{bmatrix}$ (f) $\begin{bmatrix} 3 & 1 & 1 & 0 \\ 2 & 4 & 0 & 1 \\ 1 & 2 & 6 & 3 \\ 0 & 1 & 2 & 4 \end{bmatrix}$

(g) $\begin{bmatrix} 3 & -2 & 0 & -4 \\ -2 & 5 & 2 & 1 \\ 0 & 2 & 7 & 0 \\ -4 & 1 & 0 & 10 \end{bmatrix}$ (h) $\begin{bmatrix} 1 & 2 & 3 & 4 \\ 6 & 7 & 5 & 4 \\ 2 & 1 & 1 & 0 \\ 1 & 0 & 0 & 1 \end{bmatrix}$

23. Find the second-largest eigenvalue and the corresponding eigenvector for

(a) Prob. 22e (b) Prob. 22f
(c) Prob. 22g (d) Prob. 22h

24. Write a flow chart to find the k largest eigenvalues and corresponding eigenvectors of an n by n matrix. Use the power method and deflation.

25. Find the smallest eigenvalue of the 3 by 3 matrices in Prob. 22 by using the power method.

26. Write a flow chart to find the smallest eigenvalue and the corresponding eigenvector of the n by n matrix A.

27. Show that any two eigenvectors of a symmetric matrix are orthogonal. $X_i X_j^t = 0$. Illustrate with Prob. 22b.

28. Show how the orthogonality property of the eigenvectors or a symmetric matrix can be used to evaluate $y_{1,2}$ in Eq. (4.83).

29. Prepare a flow chart for the solution of the eigenvalues of a matrix by using the interval-halving method of Chap. 3.

30. Assume an eigenvalue of a matrix has been determined and is available as data. Construct a flow chart to find the corresponding eigenvector.

31. Solve both the following systems of equations:

(a) $\begin{bmatrix} 1.00000 & 3.00000 & 1.00000 \\ 2.00001 & 6.00002 & 2.00000 \end{bmatrix}$

(b) $\begin{bmatrix} 1.00000 & 3.00000 & 1.00000 \\ 1.99997 & 5.99990 & 2.00000 \end{bmatrix}$

Why are the two solutions so different?

References

1. Faddeeva, D. K., and V. N. Faddeeva: "Computational Methods of Linear Algebra," W. H. Freeman and Company, San Francisco, 1963.
2. Faddeeva, V. N.: "Computational Methods of Linear Algebra," Dover Publications, Inc., New York, 1959.
3. Forsythe, G. E.: Solving Linear Algebraic Equations Can Be Interesting, *Bull. Am. Math. Soc.*, vol. 59, pp. 299–329, 1953.
4. ———: Tentative Classification of Methods and Bibliography on Solving Systems of Linear Equations, in "Simultaneous Linear Equations and the Determination of Eigenvalues," National Bureau of Standards, Applied Mathematics Series, vol. 29, 1953.
5. Fox, L.: "Introduction to Numerical Linear Algebra," Oxford University Press, Fair Lawn, N.J., 1964.
6. Hildebrand, F. B.: "Introduction to Numerical Analysis," McGraw-Hill Book Company, New York, 1956.
7. Householder, A. S.: "Principles of Numerical Analysis," McGraw-Hill Book Company, New York, 1953.
8. ———: "Theory of Matrices in Numerical Analysis," Blaisdell Publishing Company, New York, 1964.

9. James, E. E.: Numerical Solutions and an Application to an Autoclave, in "Thermal Stress," edited by P. P. Benham and R. D. Hoyle, Pitman Publishing Corporation, New York, 1964.
10. Kunz, K. S.: "Numerical Analysis," McGraw-Hill Book Company, New York, 1957.
11. Pipes, Louis A.: "Matrix Methods for Engineers," Prentice-Hall, Inc., Englewood Cliffs, N.J., 1963.
12. Ralston, Anthony: "A First Course in Numerical Analysis," McGraw-Hill Book Company, New York, 1965.
13. Scarborough, J. B.: "Numerical Mathematical Analysis," 5th ed., The Johns Hopkins Press, Baltimore, 1962.

Chapter 5

Interpolation and Numerical Integration

5.1 Introduction The numerical values of many functions are tabulated for discrete values of their argument. For example, the trigonometric functions are tabulated for given values of the angle. Very often the user of tables must find the value of a function for an argument that lies between two tabulated values. The procedure for finding such an approximation is called *interpolation.* If a user wishes to find the value of a function when the argument is outside the tabulated range, it is done by using the tabular values in such a way as to extend the table and is called *extrapolation.*

The problem of interpolating from a set of tabular values or extending a table is ordinarily solved by passing a polynomial through adjacent tabulated points and then assuming that the curve represents the function on the range of interest. Normally only a straight line passing through two adjacent points is used for interpolation. But interpolation formulas using higher-degree polynomials are used for other purposes. An interpolation formula can be differentiated to find an approximate value for the derivative, or it can be integrated to find an approximate value for the integral of a function.

The first part of this chapter is concerned with finding an appropriate polynomial approximation for a function that is represented by a set of tabular values. The approximation is constructed to pass through a given set of points and is based upon the change that occurs in the function from point to point. Since the points are a finite spacing apart, the methods have come to be called *finite-difference methods.* An example of a first-order finite-difference approximation to the function $f(x)$ is the straight line between the two points A and B shown in Fig. 5.1. The equation for a straight line is written in terms of the differences $f(b) - f(a)$ and $b - a$.

$$y = f(a) + \frac{f(b) - f(a)}{b - a}(x - a) \tag{5.1}$$

The function $f(x)$ is approximated on the interval $a \leq x \leq b$ by this finite-difference formula. It can be used outside the range to approximate $f(x)$. When Eq. (5.1) is used on the interval $a \leq x \leq b$ to obtain approximate values for $f(x)$, it is used as an interpolation formula. When it is used outside the interval $a \leq x \leq b$, it is used as an extrapolation formula. The suitability of the approximation will depend upon the behavior of the function on the interval. If $f(x)$ is an almost-linear function of x, the formula will give a good approximation to $f(x)$ on $a \leq x \leq b$ and may even be used with confidence outside the range. If the function is not almost linear, a higher-degree polynomial may be required to give a suitable approximation. The higher-degree polynomial can be obtained by requiring it to pass through more points on the interval.

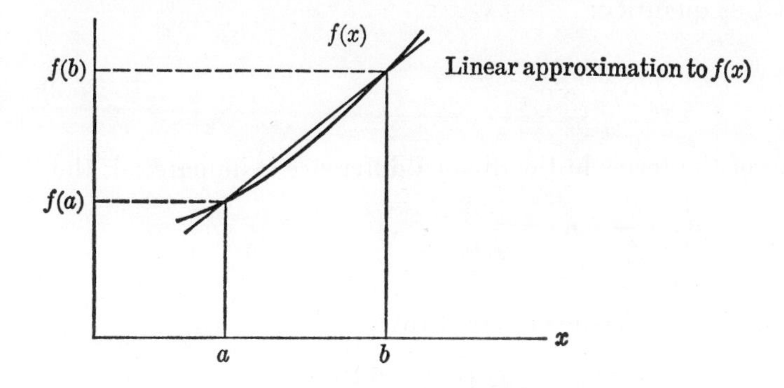

Figure 5.1 Linear approximation to a curve.

A method that provides a basis for establishing a polynomial approximation to a function is the Newton divided-difference formula. Before this method is introduced, the term *divided difference* is defined.

5.2 *Divided Differences*

The values of a function $f(x)$ are assumed to be given at $x = x_0, x_1, \ldots, x_n$. At x_0 the value of f is $f(x_0)$, at x_1 it is $f(x_1)$, and so forth. The given values of x need not be in any particular spacing, nor need they even be sequential in magnitude. However, no two of the x's can have the same value.

Assume now that one wishes to approximate $f(x)$ on the range x_0 to x_1 by a linear polynomial which passes through the points at x_0 and x_1 [see Eq. (5.1)].

$$f(x) = f(x_0) + (x - x_0)\frac{f(x_1) - f(x_0)}{x_1 - x_0} \tag{5.2}$$

The quantity

$$\frac{f(x_1) - f(x_0)}{x_1 - x_0}$$

represents an average rate of change of the function on the range $x_0 \leq x \leq x_1$. It is defined to be the first divided difference of $f(x)$ referred to the points x_0 and x_1. Square brackets are used to represent this quantity:

$$f[x_0,x_1] = \frac{f(x_0) - f(x_1)}{x_0 - x_1} \tag{5.3}$$

The order of the terms in the divided difference is immaterial; thus

$$f[x_1,x_0] = \frac{f(x_1) - f(x_0)}{x_1 - x_0} = f[x_0,x_1]$$

The first divided difference referred to x_1 and x_2 is written

$$f[x_1,x_2] = f[x_2,x_1] = \frac{f(x_1) - f(x_2)}{x_1 - x_2}$$

The first divided difference of $f(x)$ referred to x_i and x_j is

$$f[x_i,x_j] = f[x_j,x_i] = \frac{f(x_i) - f(x_j)}{x_i - x_j} \tag{5.4}$$

The second divided difference for $f(x)$ referred to the points x_i, x_j, x_k is written in terms of first divided differences.

$$f[x_i,x_j,x_k] = \frac{f[x_i,x_j] - f[x_j,x_k]}{x_i - x_k} \tag{5.5}$$

Three points are necessary to define the second divided difference. The order in which they are taken is immaterial. This can be shown by taking any permutation of x_i, x_j, x_k in the second divided difference and showing that it is equal to any other. This equivalence of any permutation of a divided difference referred to the same set of points is true for all orders of the difference. Note that the x which appears twice in the numerator does not appear in the denominator. This rule is a handy reminder of the form that the divided differences take.

The third divided difference requires four points for its definition. With reference to the points at x_0, x_1, x_2, x_3 it is

$$f[x_0,x_1,x_2,x_3] = \frac{f[x_0,x_1,x_2] - f[x_1,x_2,x_3]}{x_0 - x_3} \tag{5.6}$$

It is seen that the divided difference of order 2 is written in terms of the divided difference of order 1 and the divided difference of order 3 in terms of differences of order 2. In general, the divided difference of order n is defined in terms of the divided differences of order $n - 1$. In this way a divided difference of any order referred to an appropriate set of points can be written. The nth-order divided difference of $f(x)$ referred to the points $x_0, x_1, \ldots, x_n$ would be

$$f[x_0,x_1, \ldots ,x_n] = \frac{f[x_0,x_1, \ldots ,x_{n-1}] - f[x_1,x_2, \ldots ,x_n]}{x_0 - x_n} \tag{5.7}$$

It should be noted again that the spacing between the points is arbitrary and that the order in which they are given need not be according to magnitude. The value of x_0 may be greater than x_n, or any other noncontradictory inequalities may be satisfied except that no two x's can have the same value. The two x's that appear in the denominator are the only two that appear only once in the numerator of (5.7). All the other x's appear twice in the numerator.

5.3 Newton's Divided-difference Formula

The Newton divided-difference polynomial is constructed on the basis of the definition of the divided differences given in the previous section. Let the value of a function $f(x)$ be given at $x = x_0$, $\ldots, x_n$. Consider the divided difference of $f(x)$ referred to the two points x and x_0. Here x can take on any value on the range of approximation.

$$f[x,x_0] = \frac{f(x) - f(x_0)}{x - x_0} \tag{5.8}$$

This equation is solved for $f(x)$ to give

$$f(x) = f(x_0) + (x - x_0)f[x,x_0] \tag{5.9}$$

The divided difference $f[x,x_0]$ is replaced by using the appropriate definition of the second divided difference.

$$f[x,x_0,x_1] = \frac{f[x,x_0] - f[x_0,x_1]}{x - x_1} \tag{5.10}$$

Solving Eq. (5.10) for $f[x,x_0]$ gives

$$f[x,x_0] = f[x_0,x_1] + (x - x_1)f[x,x_0,x_1] \tag{5.11}$$

If Eq. (5.11) is substituted into Eq. (5.9), $f(x)$ takes the form

$$f(x) = f(x_0) + (x - x_0)f[x_0,x_1] + (x - x_0)(x - x_1)f[x,x_0,x_1] \tag{5.12}$$

A linear approximation for $f(x)$ which will pass through the points at x_0 and x_1 is obtained from Eq. (5.12) by truncating the last term. Call this approximation $y_1(x)$, where the subscript denotes the degree of the polynomial.

$$y_1(x) = f(x_0) + (x - x_0)f[x_0,x_1] \tag{5.13}$$

This is the Newton divided-difference formula of the first degree for $f(x)$ using the points x_0 and x_1. The second-degree formula is obtained by replacing $f[x,x_0,x_1]$ in Eq. (5.12) from the definition of the third divided difference referred to x, x_0, x_1, x_2.

$$f[x,x_0,x_1,x_2] = \frac{f[x,x_0,x_1] - f[x_0,x_1,x_2]}{x - x_2}$$

from which

$$f[x,x_0,x_1] = f[x_0,x_1,x_2] + (x - x_2)f[x,x_0,x_1,x_2]$$

Substituting into Eq. (5.12) gives

$$f(x) = f(x_0) + (x - x_0)f[x_0,x_1] + (x - x_0)(x - x_1)f[x_0,x_1,x_2] + (x - x_0)(x - x_1)(x - x_2)f[x,x_0,x_1,x_2] \tag{5.14}$$

By truncating the last term a second-degree polynomial approximation for $f(x)$ which passes through the points at x_0, x_1, x_2 is obtained.

$$y_2(x) = f(x_0) + (x - x_0)f[x_0,x_1] + (x - x_0)(x - x_1)f[x_0,x_1,x_2] \tag{5.15}$$

By a repeated replacement of the difference expression involving the arbitrary point at x a polynomial of any degree up to and including n can be constructed. The polynomial of degree $m \leq n$ will pass through the $m + 1$ points used to define it and will therefore be exactly equal to $f(x)$ at these points.

$$\begin{aligned} f(x) &= f(x_0) + (x - x_0)f[x_0,x_1] + (x - x_0)(x - x_1)f[x_0,x_1,x_2] \\ &+ \cdots + (x - x_0)(x - x_1) \cdots (x - x_{m-1})f[x_0,x_1, \ldots ,x_m] \\ &+ (x - x_0)(x - x_1) \cdots (x - x_{m-1})(x - x_m) \\ &\qquad f[x,x_0,x_1, \ldots ,x_m] \end{aligned} \tag{5.16}$$

Let $y_m(x)$ represent the mth-degree polynomial. Then

$$y_m(x) = f(x) - (x - x_0)(x - x_1) \cdots (x - x_m) \\ f[x,x_0,x_1, \ldots ,x_m] \tag{5.17}$$

It is clear that $y_m(x)$ will be equal to $f(x)$ for $x = x_0, \ldots, x_m$ since the last term in Eq. (5.17) is zero at each of these points. For values of x other than these, $y_m(x)$ and $f(x)$ will differ by $E_m(x)$, where

$$E_m(x) = (x - x_0)(x - x_1) \cdots (x - x_m)f[x,x_0,x_1, \ldots ,x_m] \tag{5.18}$$

The term $E_m(x)$ is the truncation error in the mth-degree polynomial approximation to $f(x)$. It can be computed for any x provided the mth divided difference referred to $x, x_0, x_1, \ldots, x_m$ can be found. In general it will not be possible to find this divided difference for an arbitrary x when the function is given in tabular form. However, Eq. (5.18) does provide a means for evaluating the order of approximation of the polynomial $y_m(x)$.

It is possible to find a different form for the error term in Eq. (5.18) which will prove more useful in assessing the suitability of the polynomial. The different form is in terms of the $(m + 1)$st derivative of $f(x)$, rather than the $(m + 1)$st divided difference, and it is derived below.

In Eq. (5.17) define a constant K such that for some x (not equal to x_i, $i = 0, 1, \ldots, n$) in the range of the formula, say $\bar{x}$, the error can be written as

$$f(\bar{x}) - y_m(\bar{x}) = E_m(\bar{x}) = (\bar{x} - x_0)(\bar{x} - x_1) \cdots (\bar{x} - x_m)K \tag{5.19}$$

The constant K is equal to the divided difference when x is equal to $\bar{x}$. At other values of x on the range the error term defined by Eq. (5.19) is approximate. Define a function $g(x)$ as the difference between the error in $y_m(x)$ and this estimate to the error:

$$g(x) = f(x) - y_m(x) - K(x - x_0)(x - x_1) \cdots (x - x_m) \tag{5.20}$$

The function $g(x)$ will be equal to zero for $x = \bar{x}$ and at each of the $m + 1$ points used to obtain y_m. Thus $g(x)$ will have $m + 2$ zeros in the range of interest, and the $(m + 1)$st derivative of $g(x)$ will have one zero.[4] Let this zero be at $x = \xi$.

$$g^{(m+1)}(x) = f^{(m+1)}(x) - y_m{}^{(m+1)}(x) - K \frac{d^{(m+1)}}{dx^{(m+1)}} [(x - x_0) \cdots (x - x_m)] \tag{5.21}$$

Since $y_m(x)$ is an mth-degree polynomial, its $(m + 1)$st derivative is zero. The last term in Eq. (5.20) is a polynomial of degree $m + 1$, and so its $(m + 1)$st derivative is the constant $(m + 1)!$.

The value of $g^{(m+1)}(x)$ will be zero at $x = \xi$, which is different from $x_0, x_1, \ldots, x_m$, and will be dependent upon the value of $\bar{x}$ for which K is defined. Thus

$$g^{(m+1)}(\xi) = f^{(m+1)}(\xi) - K(m + 1)! = 0 \tag{5.22a}$$

and K will be given by

$$K = \frac{1}{(m + 1)!} f^{(m+1)}(\xi) \tag{5.22b}$$

where ξ is in the interval of approximation. Since $\bar{x}$ is arbitrary, ξ can always be chosen so that the K defined in (5.22b) will make $g(x)$ in Eq. (5.20) equal to zero. The error term in the divided-difference formula can now be written in terms of the $(m + 1)$st divided difference as in Eq. (5.18) or in terms of the $(m + 1)$st derivative of $f(x)$ at ξ.

$$E_m(x) = (x - x_0)(x - x_1) \cdots (x - x_m) \frac{1}{(m + 1)!} f^{(m+1)}(\xi) \tag{5.23}$$

where ξ is some value on the interval for which the polynomial holds.

Example 5.1

To illustrate the divided-difference method the polynomial which passes through the set of four points given in Table 5.1 is constructed.

Table 5.1
Newton Divided Differences

i	x	$f(x)$	First difference	Second difference	Third difference
0	3	−5	−4	−1	⅓
1	1	3	−3	−2	
2	2	0	−1		
3	0	2			

The first-difference column in Table 5.1 is obtained from Eq. (5.4), where j is replaced by $i + 1$ and i takes on the values 0, 1, 2. The values for the second and third differences come from Eqs. (5.5) and (5.6). For example, the first entry in the second-difference column comes from the difference between the first two entries in the first-difference column divided by $x_0 - x_2$.

$$f[x_0,x_1,x_2] = \frac{f[x_0,x_1] - f[x_1,x_2]}{x_0 - x_2} = \frac{-4 - (-3)}{3 - 2} = -1$$

If the differences in Table 5.1 opposite $x = x_0 = 3$ are substituted into Eq. (5.16) with $m = 3$, the result is the following cubic equation, which passes through the four points in the table:

$$y_3(x) = \frac{x^3}{3} - 3x^2 + {}^{11}\!/_{3}x + 2$$

The Newton divided-difference formula for constructing a polynomial through a given set of points will not be used extensively in the following discussion, largely because simpler formulas using a constant spacing in the independent variable can be used in most of the applications. However, it does provide a formula for any case where the spacing is not constant. Furthermore the error term in Eq. (5.23) can be used to define the order of approximation for a polynomial. This is important in the evaluation of interpolation and integration formulas that are derived later.

5.4 *The Forward-difference Interpolation Formula*

In Sec. 5.3 the mth-degree Newton divided-difference polynomial that passes through $m + 1$ discrete points was developed. If the $m + 1$ points are ordered so that $x_i - x_{i-1} = h$ for $i = 1, 2, \ldots, m$, the coefficients in the polynomial are considerably simplified. The results of this simplification are finite-difference formulas which provide the basic approximations used for interpolation, integration, and differentiation.

Let $f(x)$ be given at $x = x_0, x_1, x_2, \ldots, x_n$, where $x_i - x_{i-1} = h$. The first divided difference referred to the points at x_0 and x_1 becomes

$$f[x_0,x_1] = \frac{f(x_1) - f(x_0)}{h} \tag{5.24}$$

The difference $f(x_1) - f(x_0)$ is defined as the forward difference of the function at x_0 and is written $\Delta f(x_0)$. Equation (5.24) then becomes

$$f[x_0,x_1] = \frac{1}{h}\,\Delta f(x_0)$$

The first divided difference referred to x_i and x_{i+1} is

$$f[x_i,x_{i+1}] = \frac{1}{h}\,\Delta f(x_i) \tag{5.25}$$

The second divided difference referred to x_0, x_1, x_2 is written in terms of the first divided difference:

$$f[x_0,x_1,x_2] = \frac{f[x_0,x_1] - f[x_1,x_2]}{x_0 - x_2} = \frac{1}{2h}\,(f[x_1,x_2] - f[x_0,x_1])$$

Replacing the first divided difference with the forward difference gives

$$f[x_0,x_1,x_2] = \frac{1}{2h^2}\,[\Delta f(x_1) - \Delta f(x_0)] = \frac{1}{2h^2}\,\Delta^2 f(x_0) \tag{5.26}$$

where

$$\Delta^2 f(x_0) = \Delta f(x_1) - \Delta f(x_0) \tag{5.26a}$$

In general, the divided difference referred to the points $x_0, x_1, \ldots,$ x_m may be written in terms of the forward difference at x_0

$$f[x_0,x_1, \ldots ,x_m] = \frac{1}{m!h^m} \Delta^m f(x_0) \tag{5.27}$$

where

$$\Delta^m f(x_0) = \Delta^{m-1} f(x_1) - \Delta^{m-1} f(x_0) \tag{5.27a}$$

The forward-difference formula for the function $f(x)$ which passes through the points at $x_0, x_1, \ldots, x_m$, where $x_i - x_{i-1} = h$, is obtained by replacing the divided differences in Eq. (5.16) by their equivalent expressions in terms of forward differences.

$$f(x) = f(x_0) + \frac{x - x_0}{h} \Delta f(x_0) + \frac{(x - x_0)(x - x_1)}{2h^2} \Delta^2 f(x_0) + \cdots$$
$$+ \frac{(x - x_0) \cdots (x - x_{m-1})}{m!h^m} \Delta^m f(x_0) + E_m(x) \tag{5.28}$$

A further simplification in the formula is possible by replacing x with $x_0 + \alpha h$, where α represents the number of spaces that x is removed from x_0. Then

$$\begin{aligned} x - x_0 &= \alpha h \\ x - x_1 &= (\alpha - 1)h \\ &\cdots\cdots\cdots \\ x - x_i &= (\alpha - i)h \end{aligned} \tag{5.28a}$$

and

$$f(x) = f(x_0 + \alpha h) = f(x_0) + \alpha \, \Delta f(x_0) + \frac{\alpha(\alpha - 1)}{2!} \Delta^2 f(x_0)$$
$$+ \cdots + \frac{\alpha(\alpha - 1)(\alpha - 2) \cdots (\alpha - m + 1)}{m!} \Delta^m f(x_0) + E_m(\alpha) \tag{5.29}$$

$$E_m(\alpha) = \frac{\alpha(\alpha - 1) \cdots (\alpha - m)}{(m + 1)!} h^{m+1} f^{(m+1)}(\xi) \qquad x_0 \le \xi \le x_m \tag{5.30}$$

Equation (5.29) is the mth-degree forward-difference approximation to $f(x)$, and Eq. (5.30) is the formula for the truncation error in this approximation.

Example 5.2

The application of formula (5.29) for approximating a function is illustrated by finding the second-degree polynomial for $\sin x$ on the interval $0 \leq x \leq \pi$. The three points used for the approximation are at $x = 0$, $\pi/2$, and π (see Table 5.2).

Table 5.2
Forward Differences

i	x	$f(x) = \sin x$	$\Delta f(x)$	$\Delta^2 f(x)$
0	0	0	1	2
1	$\pi/2$	1	-1	
2	π	0		

The forward differences are computed according to the definition given in Eq. (5.27). Note that the first forward differences are obtained from the tabular values for $f(x)$ and the second forward difference from the values for $\Delta f(x)$. The forward differences at $x = x_0$ are in the first row of the table opposite the value of $x = 0$. These differences are substituted into Eq. (5.29) for the second-degree forward-difference formula.

$$f(x_0 + \alpha h) = f(x_0) + \alpha\,\Delta f(x_0) + \tfrac{1}{2}\alpha(\alpha - 1)\,\Delta^2 f(x_0) + E_2(\alpha)$$

$$y_2(\alpha) = \alpha - \alpha(\alpha - 1) = \alpha(2 - \alpha)$$

This formula gives the exact value for the function at $\alpha = 0$, 1, and 2. These values for α correspond to values of $x = 0$, $\pi/2$, and π, respectively.

The truncation error in the approximation is given by Eq. (5.30).

$$E_2(\alpha) = \frac{\alpha(\alpha - 1)(\alpha - 2)}{6}\,h^3 f'''(\xi) = \frac{\pi^3}{48}\,\alpha(\alpha - 1)(\alpha - 2)f'''(\xi)$$

A specific value for the error cannot be obtained from this equation.

However, since the maximum value for $f'''(\xi)$ on the interval is 1, it is possible to find an upper bound for $E_2(\alpha)$.

$$|E_2(\alpha)| \leq \left| \frac{\pi^3}{48} \alpha(\alpha - 1)(\alpha - 2) \right| \qquad \text{for any } \alpha \text{ on the range}$$

Generally it will not be feasible to find a bound for the derivative term in the error. In this event only the order of the error can be given.

5.5 The Backward-difference Interpolation Formula

When the points used for Newton's divided-difference formula in Eq. (5.15) are given at $x = x_0 - mh, x_0 - (m - 1)h, \ldots, x_0$, the approximating polynomial for $f(x)$ can be written in terms of backward differences at $x = x_0$. From the definition of divided differences,

$$f[x_0, x_0 - h] = \frac{f(x_0) - f(x_0 - h)}{h}$$

The term $f(x_0) - f(x_0 - h)$ is defined as the first backward difference of $f(x)$ at $x = x_0$ and is designated $\underline{\Delta} f(x_0)$.

$$\underline{\Delta} f(x_0) = f(x_0) - f(x_0 - h) \tag{5.31}$$

The divided difference referred to x_0, $x_0 - h$, $x_0 - 2h$ is given by

$$f[x_0, x_0 - h, x_0 - 2h] = \frac{f[x_0, x_0 - h] - f[x_0 - h, x_0 - 2h]}{2h}$$

$$= \frac{1}{2h^2} [\underline{\Delta} f(x_0) - \underline{\Delta} f(x_0 - h)]$$

The second backward difference at $x = x_0$ is

$$\underline{\Delta}^2 f(x_0) = \underline{\Delta} f(x_0) - \underline{\Delta} f(x_0 - h) \tag{5.32}$$

Thus

$$f[x_0, x_0 - h, x_0 - 2h] = \frac{1}{2h^2} \underline{\Delta}^2 f(x_0)$$

The mth-order backward difference at $x = x_0$ is related to the divided difference referred to the points at $x = x_0, x_0 - h, \ldots,$

$x_0 - mh$, as follows:

$$f[x_0, x_0 - h, \ldots, x_0 - mh] = \frac{1}{m!h^m} \underline{\Delta}^m f(x_0) \tag{5.33}$$

where

$$\underline{\Delta}^m f(x_0) = \underline{\Delta}^{m-1} f(x_0) - \underline{\Delta}^{m-1} f(x_0 - h) \tag{5.34}$$

By substituting for the divided differences the equivalent expressions obtained above and by replacing x by $x_0 + \alpha h$ in Eq. (5.16), the following backward-difference formula for $f(x)$ is obtained.

$$\begin{aligned} f(x) = f(x_0 + \alpha h) &= f(x_0) + \alpha \underline{\Delta} f(x_0) \\ &\quad + \frac{\alpha(\alpha+1)}{2!} \underline{\Delta}^2 f(x_0) + \cdots \\ &\quad + \frac{\alpha(\alpha+1)\cdots(\alpha+m-1)}{m!} \underline{\Delta}^m f(x_0) + E_m(\alpha) \end{aligned} \tag{5.35}$$

where

$$E_m(\alpha) = \frac{\alpha(\alpha+1)\cdots(\alpha+m)}{(m+1)!} h^{m+1} f^{(m+1)}(\xi) \qquad x_0 - mh \le \xi \le x_0 \tag{5.36}$$

Equation (5.35) is the backward-difference polynomial for $f(x)$, and Eq. (5.36) is the truncation error in the approximation when the polynomial is of mth degree.

Example 5.3

To illustrate the backward-difference formula the third-degree polynomial which passes through the four points tabulated in Table 5.3 is found. The backward-difference table is constructed on the

Table 5.3
Backward Differences

i	x	$f(x)$	$\underline{\Delta} f(x)$	$\underline{\Delta}^2 f(x)$	$\underline{\Delta}^3 f(x)$
3	0	1			
2	1	0	−1		
1	2	−2	−2	−1	
0	3	1	3	5	6

basis of the definitions for the backward differences. For example, the first backward difference at $x = 3$ is defined as $f(3) - f(2)$, and its value is entered under $\underline{\Delta} f(x)$ opposite $x = x_0 = 3$. Its value is 3. The second divided difference at $x = x_0 - h = 2$ is defined as $\underline{\Delta} f(2) - \underline{\Delta} f(1)$ and is equal to -1.

From the results in Table 5.3 the third-degree backward-difference formula for $f(x)$ at $x = 3$ is

$$\begin{aligned} f(x) = f(x_0 + \alpha h) &\doteq f(x_0) + \alpha \, \underline{\Delta} f(x_0) + \tfrac{1}{2}\alpha(\alpha + 1) \, \underline{\Delta}^2 f(x_0) \\ &\qquad + \tfrac{1}{6}\alpha(\alpha + 1)(\alpha + 2) \, \underline{\Delta}^3 f(x_0) \\ &= 1 + 3\alpha + \tfrac{5}{2}\alpha(\alpha + 1) \\ &\qquad + \tfrac{6}{6}\alpha(\alpha + 1)(\alpha + 2) \\ &= \alpha^3 + \tfrac{11}{2}\alpha^2 + \tfrac{15}{2}\alpha + 1 \end{aligned}$$

The truncation error in the approximation to $f(x)$ is given by Eq. (5.36).

$$E_3(\alpha) = \frac{h^4}{4!} \alpha(\alpha + 1)(\alpha + 2)(\alpha + 3) f^{\text{IV}}(\xi)$$

In this case it is not possible to find bounds on the error since nothing is known about $f^{\text{IV}}(\xi)$. The error is of $O(h^4)$, and this gives some indications of relative improvement of the formula when h is changed.

5.6 *Approximate Integration Formulas*

The integration of a function $f(x)$ in terms of elementary functions can be made only for special forms of $f(x)$. If $f(x)$ is not integrable in this sense, it will generally be necessary to resort to some numerical method to find an estimate for the integral. A general procedure for finding an approximate numerical value for an integral is introduced in this section. The basis for the method is to replace the integrand by the backward-difference polynomial developed in the last section. The integration of the polynomial can then be made in a straightforward manner.

Consider the integration of a function $f(x)$ on the interval x_{i-r} to x_{i+s}, where $x_{i-r} = x_i - rh$, $x_{i+s} = x_i + sh$. The limits of integration are defined by the integers i, r, and s, so that the integral is a function of i, r and s.

$$I(i,r,s) = \int_{x_{i-r}}^{x_{i+s}} f(x)\,dx \tag{5.37}$$

To obtain an approximation to the integral the integrand is replaced by the backward-difference polynomial which approximates $f(x)$ on the interval $x_{i-r} \leq x \leq x_{i+s}$. The appropriate backward-difference polynomial is obtained from Eq. (5.35) by replacing x_0 by x_i.

$$\begin{aligned} f(x) = f(x_i + \alpha h) &= f(x_i) + \alpha\,\underline{\Delta} f(x_i) + \frac{\alpha(\alpha+1)}{2!}\,\underline{\Delta}^2 f(x_i) \\ &\quad + \frac{\alpha(\alpha+1)(\alpha+2)}{3!}\,\underline{\Delta}^3 f(x_i) + \cdots \\ &\quad + \frac{\alpha(\alpha+1)\cdots(\alpha+m-1)}{m!}\,\underline{\Delta}^m f(x_i) + E_m(\alpha) \end{aligned} \tag{5.38}$$

where

$$E_m(\alpha) = \frac{h^{m+1}}{(m+1)!}\,\alpha(\alpha+1)\cdots(\alpha+m)f^{(m+1)}(\xi) \qquad x_{i-r} \leq \xi \leq x_{i+s} \tag{5.39}$$

If the polynomial for $f(x)$ in Eq. (5.38) is substituted into Eq. (5.37), and if dx is replaced by its equivalent in terms of $d\alpha$, the integration can be carried out. From

$$x = x_i + \alpha h$$

we get

$$dx = h\,d\alpha$$

The limits of integration in terms of α come from the equalities $x_{i+s} = x_i + sh$ and $x_{i-r} = x_i - rh$. When $x = x_{i+s}$, $\alpha = s$, and when $x = x_{i-r}$, $\alpha = -r$.

$$\begin{aligned} I(i,r,s) = h\int_{-r}^{s}\Big[& f(x_i) + \alpha\,\underline{\Delta} f(x_i) + \cdots \\ & + \frac{\alpha(\alpha+1)\cdots(\alpha+m-1)}{m!}\,\underline{\Delta}^m f(x_i)\Big]\,d\alpha \\ & + h\int_{-r}^{s} E_m(\alpha)\,d\alpha \end{aligned} \tag{5.40}$$

Performing the integration and collecting coefficients on the backward differences gives the following important result.

$$I(i,r,s) = h\left[\alpha f(x_i) + \frac{\alpha^2}{2}\,\Delta f(x_i) + \frac{\alpha^2}{12}(2\alpha + 3)\,\Delta^2 f(x_i) + \frac{\alpha^2}{24}(\alpha + 2)^2\,\Delta^3 f(x_i) + \frac{\alpha^2}{720}(6\alpha^3 + 45\alpha^2 + 110\alpha + 90)\,\Delta^4 f(x_i) + \cdots\right]_{\alpha=-r}^{\alpha=s} + \frac{h^{m+2}}{(m+1)!}\int_{-r}^{s} \alpha(\alpha+1)\cdots(\alpha+m)f^{(m+1)}(\xi)\,d\alpha \tag{5.41}$$

where m is the highest-order difference that is retained in the formula. The integral term in Eq. (5.41) is the truncation error in the integration formula.

The backward-difference approximation to $f(x)$ uses the points at $x_i, x_{i-1}, x_{i-2}, \ldots, x_{i-m}$. If the limits of integration do not go outside the interval x_{i-m} to x_i, the formula is said to be a closed-end integration formula. If the limits of integration extend beyond the points used to make up the difference polynomial, the integration formula is an open-end formula. Both types are used for numerical integration.

5.7 *Trapezoidal Formula for Integration*

Many integration formulas can be obtained from Eq. (5.41) by varying the number of terms used in the approximation and by varying the limits of integration. An important result is obtained when two terms are used and the limits of integration are $r = 1$ and $s = 0$. This is equivalent to using a straight-line approximation for $f(x)$ between x_{i-1} and x_i and gives

$$I(i,1,0) = \int_{x_{i-1}}^{x_i} f(x)\,dx = \frac{h}{2}[f(x_i) + f(x_{i-1})] + E_T \tag{5.42}$$

The limits of integration are within the interpolation range of the approximation to $f(x)$, so that the result is a closed-end integration formula. The truncation error for this approximation is given by

$$E_T = \frac{h^3}{2}\int_{-1}^{0} \alpha(\alpha+1)f''(\xi)\,d\alpha \doteq -\frac{h^3}{12}f''(\xi) \tag{5.43}$$

The formula in (5.42) is the trapezoidal rule for integration between x_{i-1} and x_i. In Fig. 5.2 it can be seen that Eq. (5.42) gives the area enclosed by the trapezoid under the straight-line approximation to the curve between x_{i-1} and x_i. The error in formula (5.42) is $O(h^3)$, so that by reducing the spacing in the interpolation

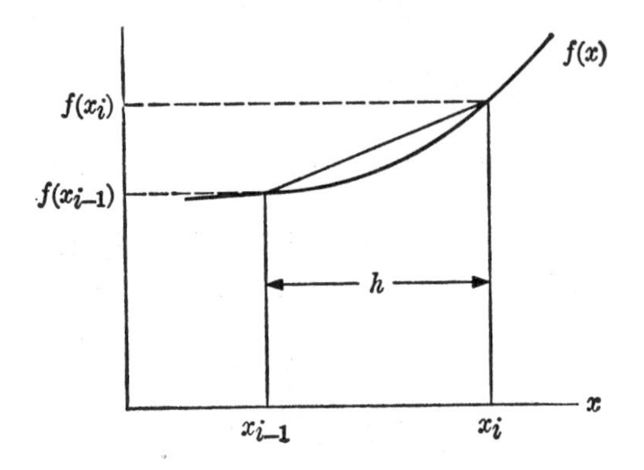

Figure 5.2 Trapezoidal rule for integration on x_{i-1} to x_i.

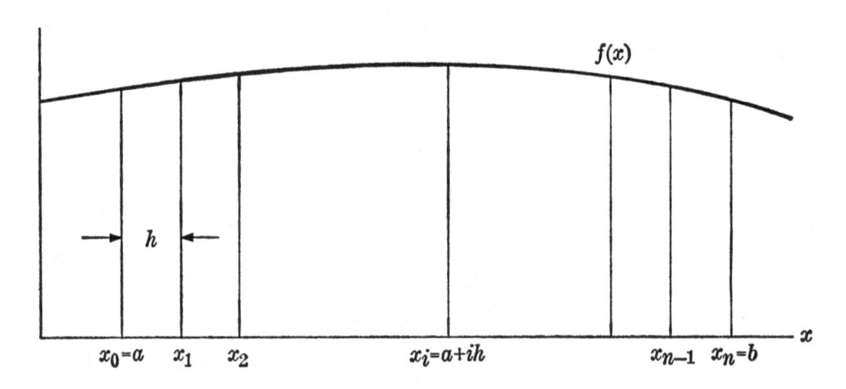

Figure 5.3 Trapezoidal rule for integration on a to b.

formula the error is reduced significantly. For example, if the spacing h is reduced by one-half, the error is reduced to approximately one-eighth its previous value.

The application of the trapezoidal rule for integration is illustrated by integrating the function $f(x)$ on the interval from $x = a$ to $x = b$. The spacing on a to b is divided into n equal spaces by the lines at $x = a,\ a + h,\ a + 2h,\ \ldots,\ a + nh = b$, and the value of $f(x)$ at each of these points is determined (see Fig. 5.3). Let $x_i = a + ih$. By the trapezoidal formula the integral of $f(x)$ from a to $a + h$ is given by

$$\int_{x_0}^{x_1} f(x)\,dx = \int_a^{a+h} f(x)\,dx = \frac{h}{2}[f(a) + f(a+h)] - \tfrac{1}{12}h^3 f''(\xi_1)$$

Similarly the integral from $a + h$ to $a + 2h$ is given by

$$\int_{x_1}^{x_2} f(x)\,dx = \int_{a+h}^{a+2h} f(x)\,dx = \frac{h}{2}[f(a+h) + f(a+2h)] - \tfrac{1}{12}h^3 f''(\xi_2)$$

and for the ith space

$$\int_{x_{i-1}}^{x_i} f(x)\,dx = \frac{h}{2}[f(x_{i-1}) + f(x_i)] - \tfrac{1}{12}h^3 f''(\xi_i)$$

The integral over the whole range a to b is the sum of the integrals over the parts.

$$\begin{aligned}\int_a^b f(x)\,dx &= \sum_{i=1}^{n} \int_{x_{i-1}}^{x_i} f(x)\,dx = \sum_{i=1}^{n} \frac{h}{2}[f(x_{i-1}) + f(x_i)] - \tfrac{1}{12}h^3 \sum_{i=1}^{n} f''(\xi_i) \\ &= \frac{h}{2}[f(a) + 2f(a+h) + \cdots + 2f(a+ih) + \cdots + 2f(a+nh-h) + f(b)] + E_T \qquad (5.44)\end{aligned}$$

The error term in Eq. (5.44) can be expressed in terms of an average value for $f''(x)$. Let this be $\bar{f}''$.

$$E_T = -\tfrac{1}{12}h^3 \sum_{i=1}^{n} f''(\xi_i) = -\tfrac{1}{12}h^3 n\bar{f}'' = -\tfrac{1}{12}(nh)h^2\bar{f}''$$

Using the relation $b = x_n = a + nh$,

$$E_T = -\tfrac{1}{12}(b - a)h^2\bar{f}'' \qquad (5.45)$$

Equation (5.45) shows the error term on the whol interval a to b to be of $O(h^2)$. While reducing the spacing h by one-half reduces the error in each interval by one-eighth, there are twice as many intervals, and so the error in Eq. (5.44) is reduced by only one-fourth.

An example is solved using the trapezoidal rule for integration, and then a flow chart is constructed for trapezoidal integration.

Example 5.4

Find the integral of $f(x) = x \sin x$ on the interval $0 \leq x \leq \pi$. Table 5.4 is a table of values for the integrand for values of x which vary by $\pi/6$.

Table 5.4
Tabular Values of
$f(x) = x \sin x$

i	x	$f(x) = x \sin x$
0	0	0
1	$\pi/6$	$\pi/12$
2	$\pi/3$	$\pi/2\sqrt{3}$
3	$\pi/2$	$\pi/2$
4	$2\pi/3$	$\pi/\sqrt{3}$
5	$5\pi/6$	$5\pi/12$
6	π	0

The values of $f(x)$ listed in Table 5.4 are substituted into Eq. (5.44).

$$\int_0^\pi x \sin x \, dx = \frac{\pi}{12}\left(\frac{2\pi}{12} + \frac{\pi}{\sqrt{3}} + \pi + \frac{2\pi}{\sqrt{3}} + \frac{10\pi}{12}\right)$$

$$= \frac{\pi^2}{12}(2 + \sqrt{3}) = 3.0694$$

The exact value for the integral is π. The error in the approximation is 0.0722, which is 2.3 percent of the exact value. If every other point in the table above is used, i.e., $h = \pi/3$, the approximation is given by the expression

$$\int_0^\pi x \sin x \, dx = \frac{\pi}{6}\left(\frac{\pi}{\sqrt{3}} + \frac{2\pi}{\sqrt{3}}\right) = \frac{\pi^2}{2\sqrt{3}} = 2.8492$$

The error in this approximation is 0.2924, which is 9.3 percent of the exact value. The error in the latter case for $h = \pi/3$ is almost exactly four times the error when $h = \pi/6$.

Since the truncation error in the trapezoidal formula is $O(h^2)$, it is expected that the ratio of the two errors in the integral will have approximately the same value as the square of the ratio of the corresponding values for h. Thus, if E_1 and h_1 correspond to the error and

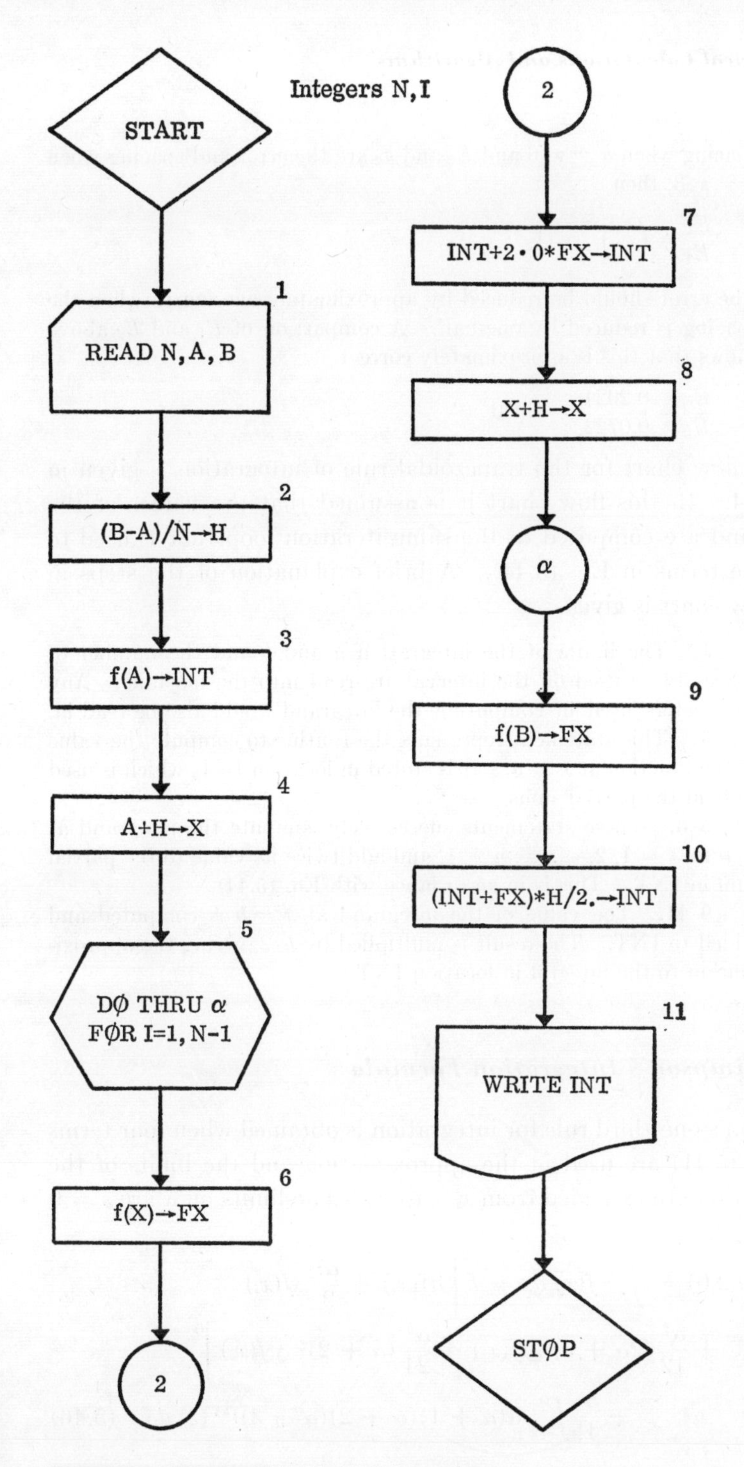

Figure 5.4 Trapezoidal rule for integration.

spacing when $h = \pi/6$ and E_2 and h_2 are the error and spacing when $h = \pi/3$, then

$$\frac{E_1}{E_2} \doteq \left(\frac{h_1}{h_2}\right)^2 = (1/2)^2 = 1/4$$

The error should be reduced by approximately one-fourth when the spacing is reduced by one-half. A comparison of E_1 and E_2 above shows that this is approximately correct.

$$\frac{E_2}{E_1} = \frac{0.2924}{0.0722} = 4.050$$

A flow chart for the trapezoidal rule of integration is given in Fig. 5.4. In this flow chart it is assumed that the values of the integrand are computed in the same iteration loop that is used to add the terms in Eq. (5.44). A brief explanation of the steps in the flow chart is given.

1. The limits of the integration a and b and the number of spaces to be used on the interval are read into the machine. Any parameters used in computing the integrand would also be read in.

3. This statement represents the routine to compute the value of the function at $x = a$. It is stored in location INT, which is used to hold the partial sums.

5–8. These statements successively compute the integrand at $a + ih$, $i = 1, 2, \ldots, n - 1$, and add twice its value to the partial sum in INT. This is in accordance with Eq. (5.44).

9–11. The value of the integrand at $x = b$ is computed and added to INT. The result is multiplied by $h/2$, giving the approximation to the integral in location INT.

5.8 Simpson's Integration Formula

Simpson's one-third rule for integration is obtained when four terms of Eq. (5.41) are used in the approximation and the limits of the integration are extended from x_{i-2} to x_i. The limits on α are $s = 0$ and $r = 2$.

$$I(i,2,0) = \int_{x_{i-2}}^{x_i} f(x)\,dx = h\left[\alpha f(x_i) + \frac{\alpha^2}{2}\,\Delta f(x_i) + \frac{\alpha^2}{12}(2\alpha + 3)\,\Delta^2 f(x_i) + \frac{\alpha^2}{24}(\alpha + 2)^2\,\Delta^3 f(x_i)\right]_{\alpha=-2}^{0} + \frac{h^5}{4!}\int_{-2}^{0} \alpha(\alpha + 1)(\alpha + 2)(\alpha + 3) f^{\mathrm{IV}}(\xi)\,d\alpha \qquad (5.46)$$

When the limits on α are substituted into the expression in brackets, it is noted that the coefficient on the third difference is zero. This simplifies considerably the integration formula while retaining the error term for the formula with four terms. Substitution for α in Eq. (5.46) gives

$$\int_{x_{i-2}}^{x_i} f(x)\,dx = \frac{h}{3}\,[f(x_i) + 4f(x_{i-1}) + f(x_{i-2})] + E_s \tag{5.47}$$

The error is obtained by carrying out the integration of the error term in Eq. (5.46):

$$E_s = -\tfrac{1}{90}h^5 f^{\mathrm{IV}}(\xi) \tag{5.48}$$

The integration formula in Eq. (5.47) is based on approximating $f(x)$ on the range $x_{i-2} \leq x \leq x_i$ by the second-degree polynomial that is equal to $f(x)$ at x_{i-2}, x_{i-1}, x_i (see Fig. 5.5).

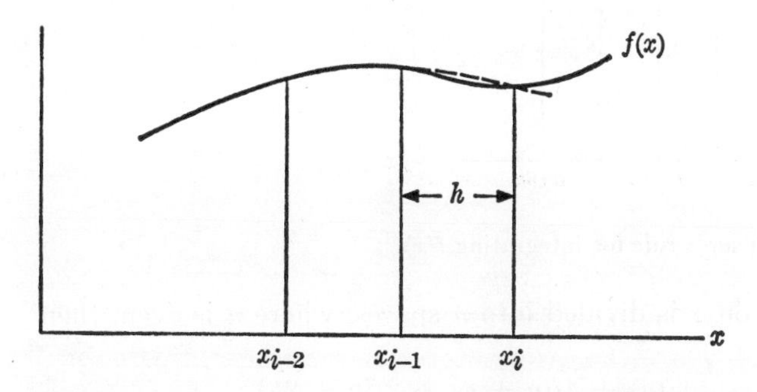

Figure 5.5 Simpson's one-third rule for integration on x_{i-2} to x_i.

Application of Simpson's one-third rule integration formula over the range a to b is made by dividing the interval into an even number of equal spaces, as shown in Fig. 5.6, and then using Eq. (5.47) to find the integrand over consecutive pairs of the spaces. The integral approximation for the region a to $a + 2h$ is the area under the second-degree curve marked (1) in Fig. 5.6.

$$\int_a^{a+2h} f(x)\,dx = \frac{h}{3}\,[f(a) + 4f(a+h) + f(a+2h)] - \tfrac{1}{90}h^5 f^{\mathrm{IV}}(\xi_2) \tag{5.49}$$

The integral from $a + 2h$ to $a + 4h$ is the area under the second-degree curve marked (2) in Fig. 5.6.

$$\int_{a+2h}^{a+4h} f(x)\,dx = \frac{h}{3}\,[f(a+2h) + 4f(a+3h) + f(a+4h)] - \tfrac{1}{90}h^5 f^{\mathrm{IV}}(\xi_4) \qquad (5.50)$$

An approximation to the integral of $f(x)$ on the interval a to b when $h = (b - a)/4$ is given by the sum of Eqs. (5.49) and (5.50).

$$\int_a^b f(x)\,dx = \frac{h}{3}\,[f(a) + 4f(a+h) + 2f(a+2h) + 4f(a+3h) + f(a+4h)] - \tfrac{1}{90}h^5[f^{\mathrm{IV}}(\xi_2) + f^{\mathrm{IV}}(\xi_4)] \qquad (5.51)$$

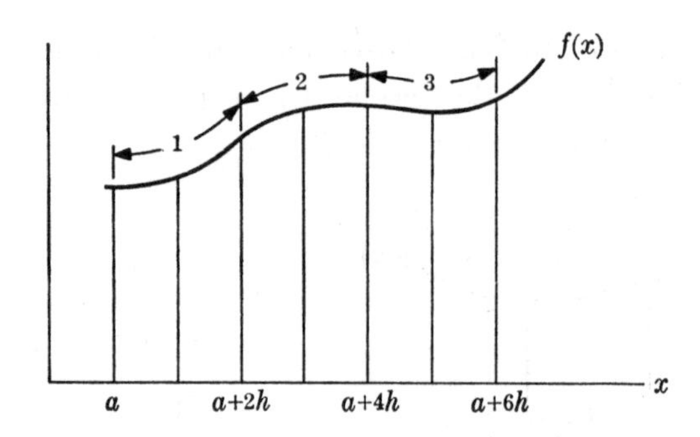

Figure 5.6 Simpson's rule for integrating $f(x)$.

If the interval on x is divided into n spaces, where n is even, then

$$\int_a^b f(x)\,dx = \frac{h}{3}\,[f(a) + 4f(a+h) + 2f(a+2h) + 4f(a+3h) + \cdots + 4f(a+nh-h) + f(a+nh)] - \tfrac{1}{90}h^5 \sum_{i=1}^{n/2} f^{\mathrm{IV}}(\xi_{2i}) \qquad (5.52)$$

If an average value for $f^{\mathrm{IV}}(x)$, $\bar{f}^{\mathrm{IV}}$ is used, the error term in Eq. (5.52) becomes

$$E_5 = -\tfrac{1}{90}h^5\bar{f}^{\mathrm{IV}} \sum_{i=1}^{n/2} 1 = -\tfrac{1}{180}h^4(b-a)\bar{f}^{\mathrm{IV}} \qquad (5.53)$$

Thus the error term in Simpson's one-third rule is $O(h^4)$. Reducing the spacing by one-half should reduce the error by about one-sixteenth, a considerable improvement over the trapezoidal rule.

Example 5.5

The function tabulated in Table 5.4 is integrated twice by Simpson's one-third rule, first using every third point and second using every point. The results are for $h_1 = \pi/2$:

$$\int_0^\pi f(x)\,dx = \frac{\pi}{6}\left(0 + 4\frac{\pi}{2} + 0\right) = \frac{\pi^2}{3} = 3.2899$$

for $h_2 = \pi/6$:

$$\int_0^\pi f(x)\,dx = \frac{\pi}{18}\left(0 + 4\frac{\pi}{12} + 2\frac{\pi}{2\sqrt{3}} + 4\frac{\pi}{2} + 2\frac{\pi}{\sqrt{3}} + 4\frac{5\pi}{12} + 0\right)$$

$$= \frac{\pi^2}{18}(4 + \sqrt{3}) = 3.1429$$

The error in the first result is $E_1 = 0.1483$; in the second, $E_2 = 0.0013$. Thus,

$$\frac{E_1}{E_2} = 114$$

The fourth power of the ratio of the spacing is

$$\left(\frac{h_1}{h_2}\right)^4 = 3^4 = 81$$

The difference in these two ratios illustrates the fact that the order of the error term, viz., h^4, will give only an estimate of the improvement in the solution.

A flow chart for Simpson's rule is given in Fig. 5.7. In it the approximate value for the integral is found, first for n spaces on the interval of integration, then for $2n$ spaces, then for $4n$ spaces, etc. For any evaluation of the integral the number of spaces is twice the number of spaces for the preceding result. The criterion for stopping the iteration is the value of the difference between two successive results.

5.9 *Other Closed-end Integration Formulas*

Higher-order closed-end integration formulas can be constructed from Eq. (5.41) by increasing the number of terms in the interpolation polynomial for $f(x)$. The interval of integration can be varied

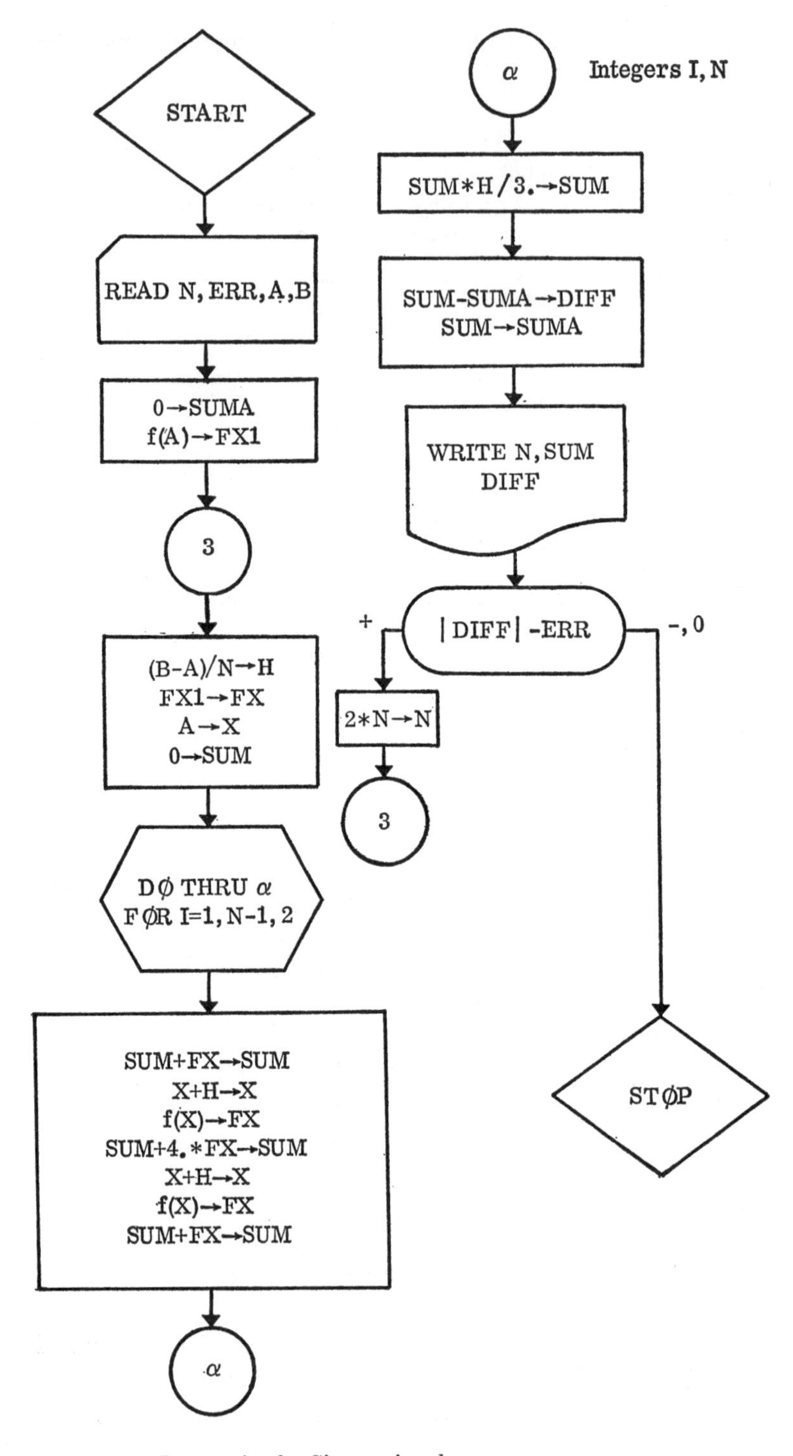

Figure 5.7 Integration by Simpson's rule.

to give several different formulas having the same order of truncation error. Such formulas are not used extensively for evaluating definite integrals but do find application in the solution of differential equations.

An integration formula that is used in the solution of initial-value problems in the next chapter is obtained by setting $s = 0$, $r = 1$ and including $m + 1$ terms in the approximation to the integral. This is equivalent to passing an mth-degree interpolation polynomial through the points at $x_i, x_{i-1}, \ldots, x_{i-m}$ and then integrating the approximation that is obtained from x_{i-1} to x_i. The result, given in Eq. (5.54), includes terms through the fifth difference $(m = 5)$.

$$I(i,1,0) = \int_{x_{i-1}}^{x_i} f(x)\,dx = h[f(x_i) - \tfrac{1}{2}\Delta f(x_i) - \tfrac{1}{12}\Delta^2 f(x_i) - \tfrac{1}{24}\Delta^3 f(x_i) - \tfrac{19}{720}\Delta^4 f(x_i) - \tfrac{3}{160}\Delta^5 f(x_i) - \cdots] + E_m \quad (5.54)$$

When the formula in Eq. (5.54) is truncated after the mth difference, the error is given by E_m, where

$$E_m = \frac{h^{m+2}}{(m+1)!}\int_{-1}^{0} \alpha(\alpha+1)\cdots(\alpha+m)f^{(m+1)}(\xi)\,d\alpha \quad (5.55)$$

Equation (5.54) is not ordinarily used for purposes of evaluating definite integrals but is used extensively in the solution of differential equations. This will be discussed in the next chapter, and reference will be made to Eq. (5.54) and its associated error term (5.55).

5.10 Open-end Integration Formulas

The open-end integration formulas are obtained when the interval of integration extends beyond the range of interpolation. If in the backward-difference formula in Eq. (5.41) a value greater than zero is assigned to s, or if the number of terms that is retained in the interpolating polynomial is equal to or less than r, the integration will extend outside the interpolation range of the backward-difference polynomial. Open-end integration formulas are not ordinarily used in the numerical evaluation of an integral, but they do have an important application in the solution of differential equations. One general form of an open-end formula that will be used in the solution of differential equations is developed here for reference.

An estimate to the integral on the range x_i to x_{i+1} can be obtained to an order of h^{m+2} by using terms up to and including the mth difference in Eq. (5.41). If $s = 1$ and $r = 0$, the result through the fifth difference is

$$I(i,0,1) = \int_{x_i}^{x_{i+1}} f(x)\,dx = h[f(x_i) + \tfrac{1}{2}\underline{\Delta} f(x_i) + \tfrac{5}{12}\underline{\Delta}^2 f(x_i) + \tfrac{3}{8}\underline{\Delta}^3 f(x_i) + \tfrac{251}{720}\underline{\Delta}^4 f(x_i) + \tfrac{95}{288}\underline{\Delta}^5 f(x_i) + \cdots] + E_m \tag{5.56}$$

The error in this approximation when terms through the mth difference are used is given by E_m.

$$E_m(\alpha) = \frac{h^{m+2}}{(m+1)!}\int_0^1 \alpha(\alpha+1)\cdots(\alpha+m)f^{(m+1)}(\xi)\,d\alpha \tag{5.57}$$

This formula has the important feature that the integration is extended beyond the range of values used in the approximation for $f(x)$ while at the same time a truncation error of arbitrary order can be obtained.

5.11 Comparison of Integration Formulas

The integration formulas presented here have all been derived from the backward-difference interpolation formula for $f(x)$. Had the forward-difference interpolation formula been used, an integration formula similar to Eq. (5.41) would have been obtained. Upon substitution of the appropriate values for s and r and a proper choice of the number of terms in the forward-difference formula exactly the same results could be obtained for the closed-end formulas. The open-end formulas with an arbitrary number of terms could also be derived with forward differences but they would be in terms of an arbitrary number of points ahead of x_i. Although the form for the open-end integration formula using forward differences could have the same accuracy, it would not be suitable for use in the next chapter.

For virtually all applications of numerical integration the closed-end formulas are used because of their greater accuracy and simplicity. Of the closed-end formulas the trapezoidal rule and Simpson's one-third rule are by far the most popular. The former is adequate for cases where the function is slowly varying on the interval of approximation. If the function is changing rapidly, it can certainly be better approximated by a second-degree curve, and Simpson's one-third rule will give better results. Improvement in the approximation in each case is made by reducing h, and the improvement is faster with Simpson's rule. Of these two formulas Simpson's rule gives a better approximation for a given spacing, and the improvement in the approximation is faster as h is reduced. Since Simpson's formula is almost as easily programmed as the trapezoidal rule, it normally is used. However, when a high order of accuracy is not required or when the integrand is slowly varying, the trapezoidal formula may be adequate.

A word of caution is added here in regard to reducing the spacing on a given interval. Although the truncation-error term can be arbitrarily reduced, a very large number of spaces can admit round-off errors that will offset the improvement made in the truncation error. The procedure suggested in the flow chart for Simpson's rule in Fig. 5.7 has been found very effective in identifying a suitable spacing.

5.12 Summary

The divided-difference polynomial of Newton is defined, and the error associated with a polynomial approximation to a function is derived. The forward and backward differences of a function are defined and illustrated with examples, and the finite-difference interpolation polynomials using forward and backward differences are derived. A number of integration formulas are obtained by integrating the backward-difference interpolation formula for $f(x)$. Some examples illustrating the application of the trapezoidal rule and Simpson's rule for integration are given.

Problems

1. Construct the second-degree-polynomial approximation to $\sin \pi x$ on the interval $0 \le x \le \frac{1}{2}$ by using each of the following sets of values for x:

(a) $x = 0, \frac{1}{4}, \frac{1}{2}$ (b) $x = 0, \frac{1}{6}, \frac{1}{2}$
(c) $x = 0, \frac{1}{3}, \frac{1}{2}$ (d) $x = 0, \frac{1}{6}, \frac{1}{3}$
(e) $x = \frac{1}{6}, \frac{1}{3}, \frac{1}{2}$

2. Find the error in each of the approximations in Prob. 1 for $x = \frac{3}{8}$.

3. Find the divided-difference formula that will pass through each of the following set of points:

(a) (2,1), (0,7), (1,2) (b) (4,0), (1,−8)
(c) (0,−10), (−1,0), (3,6), (5,10)
(d) (2,0), (0,6), (−1,2), (−3,4)
(e) (0,0), (1,1), (2,2), (4,4)
(f) (0,1), (1,−1), (2,3), (4,4)

4. Construct a flow chart to evaluate the mth divided difference of a function $y(x)$ when $m + 1$ values of the function are given.

5. Construct a flow chart to evaluate the mth-degree divided-difference polynomial that will pass through the $m + 1$ points (x_i,y_i), $i = 0, 1, 2, \ldots, m$.

6. The mth-degree polynomial for $y(x)$ is given. Construct a flow chart to find values of the function y at values of $x = x_0 + jh$, $j = 0, 1, 2, \ldots, n$.

7. Find the forward-difference polynomial that will pass through the following set of points:

(a) (1,0), (2,3) (b) (0,−2), (1,0), (2,0)
(c) (−2,4), (0,0), (2,6) (d) (−1,3), (0,0), (1,3)
(e) (−1,−3), (0,−1), (1,3), (2,0)
(f) $(-\frac{1}{2},1)$, (0,4), $(\frac{1}{2},8)$, (1,5), $(\frac{3}{2},0)$

8. Construct a flow chart to compute the mth forward difference of a tabulated function.

9. Construct a flow chart to evaluate the mth-degree forward-difference polynomial that passes through the points at $x_0 + jh$, $j = 0, 1, \ldots, m$.

10. Find the backward-difference polynomial that passes through each set of points given in Prob. 7.

11. Use the trapezoidal rule to evaluate the integral of the functions tabulated in Prob. 7.

12. Do Prob. 11 using Simpson's one-third rule.

13. Construct a flow chart to integrate $(\sin x)/x$ on the range $0 \leq x \leq \pi/2$ using Simpson's one-third rule.

14. A test for the suitability of a numerical integration can be made by reducing the spacing on the interval and then comparing the result for the reduced spacing with the previous result. Incorporate into the flow chart for the trapezoidal rule a test such that two successive integrations will be compared and if the difference is greater than some allowable error, a new computation will be made with the spacing reduced by one-half.

15. Do Prob. 13 using the test suggested in Prob. 14.

16. The speed of an object is measured every ¼ sec for 2 sec and the results tabulated.

Time, sec	0.00	0.25	0.50	0.75	1.00	1.25	1.50	1.75	2.00
Speed, ft/sec	5.0	4.9	4.7	5.2	5.6	6.1	7.0	9.0	12.00

What is the approximate distance traveled by the object in the 2 sec?

17. Write a flow chart for Prob. 16 using t sec and p intervals per second. Use (a) trapezoidal rule; (b) Simpson's rule. What restrictions must be imposed on p?

18. Construct a flow chart to find the area under an nth-degree polynomial on the interval $0 \leq x \leq 2$. (a) Use the trapezoidal rule with the test criterion suggested in Prob. 14. (b) Use Simpson's rule with the test criterion suggested in Prob. 14.

19. Construct a flow chart to find the area under each of the polynomials in Prob. 1 of Chap. 3 on the interval a to b using the trapezoidal rule and the test criterion suggested in Prob. 14.

20. Construct a flow chart to find the area under each of the polynomials in Prob. 13 of Chap. 3 on the interval a to b using Simpson's rule and the test criterion suggested in Prob. 14.

References

1. Butler, R., and E. Kerr: "An Introduction to Numerical Methods," Pitman Publishing Corporation, New York, 1962.
2. Hamming, R. W.: "Numerical Methods for Scientists and Engineers," McGraw-Hill Book Company, New York, 1962.
3. Henrici, P.: "Elements of Numerical Analysis," John Wiley & Sons, Inc., New York, 1964.
4. Hildebrand, F. B.: "Introduction to Numerical Analysis," McGraw-Hill Book Company, New York, 1956.

5. Ralston, Anthony: "A First Course in Numerical Analysis," McGraw-Hill Book Company, New York, 1965.
6. Salvadori, M. G., and M. L. Barron: "Numerical Methods in Engineering," Prentice-Hall, Inc., Englewood Cliffs, N.J., 1952.
7. Scarborough, J. B.: "Numerical Mathematical Analysis," 5th ed., The Johns Hopkins Press, Baltimore, 1962.
8. Stanton, R. G.: "Numerical Methods for Science and Engineering," Prentice-Hall, Inc., Englewood Cliffs, N.J., 1961.

Chapter 6
Initial-value Problems

6.1 Introduction The relationship between variables in many problems in science and engineering is expressed in terms of a differential equation. One of the simplest examples is the equation of the free fall of a mass m in a gravitational field g. The equation relating the displacement of the mass to the independent variable t (time) is

$$m \frac{d^2y}{dt^2} = -mg \tag{6.1}$$

The solution to Eq. (6.1) defines the displacement y as a function of the time t and is found directly by integration.

$$y = -\tfrac{1}{2}gt^2 + c_1t + c_2 \tag{6.2}$$

The constants c_1 and c_2 are constants of integration and allow freedom to fit the general solution [Eq. (6.2)] to any free-fall problem. If the values for the displacement y and the velocity dy/dt are given for some time $t = t_0$, the constants c_1 and c_2 are uniquely defined. A particular solution curve is obtained which will give unique values to y and dy/dt for all values of time.

Figure 6.1 shows a solution which passes through (t_0,y_0) with a slope equal to dy_0/dt. There is an infinite number of solutions which pass through the point (t_0,y_0), each with a different slope (see curve A). Likewise there is an infinite number of solution curves which have a given slope at $t = t_0$ but have different values for the displacement y (see curves B and C). The special curve that gives a solution to a specific free-fall problem is determined by prescribing values for both the displacement and slope at some value of time. Ordinarily this is the starting time for the problem, and the values for displacement and slope are the initial values for the variables.

Problems of the type discussed above, where the conditions at some given value of the independent variable define a unique solution, are called initial-value problems. The property that the solution for any value of the independent variable is totally dependent upon the conditions at some initial value of the variable forms the basis for a numerical solution of the initial-value problem. The numerical methods that are employed to solve this problem are developed using the first-order differential equation. They are then extended to include systems of equations and higher-order differential equations.

6.2 *The First-order Differential Equations*

The first-order differential equation can be written as some functional relationship between an independent variable x, a dependent variable y, and its derivative dy/dx.

$$F\left(x, y, \frac{dy}{dx}\right) = 0$$

It is assumed at the outset that this equation can be solved explicitly

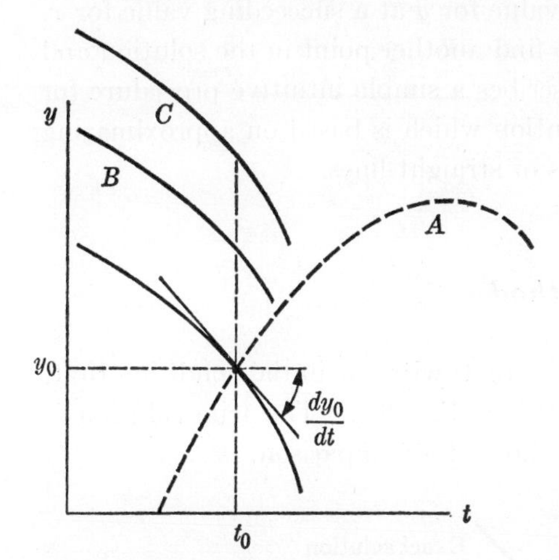

Figure 6.1 Solution curves for Eq. (6.1).

for the derivative dy/dx, so that it can always be represented in the form

$$y' = \frac{dy}{dx} = f(x,y) \tag{6.3}$$

The function $f(x,y)$ is assumed to have a unique value for any admissible value of its arguments. This restriction on the differential equation is not severe from the practical viewpoint, since physical problems generally fall into this category.

The solution to Eq. (6.3) is an explicit function of x which contains one constant of integration. Let the solution be represented by the function $g(x,c)$, where c is the constant of integration.

$$y = g(x,c) \tag{6.4}$$

Equation (6.4) represents a family of curves in the xy plane all of which satisfy Eq. (6.3). There is one and only one curve of the family which passes through any given point in the plane. Equation (6.5) defines the constant c, which, when placed in Eq. (6.4), gives the particular curve of the family that passes through (x_0,y_0) to give a unique solution to Eq. (6.3).

$$y_0 = g(x_0,c) \tag{6.5}$$

A numerical solution to the initial-value problem is based on starting at the initial point (x_0,y_0) and using the differential equation to find an approximate value for y at a succeeding value for x. This estimate is then used to find another point in the solution and so on. The next section describes a simple intuitive procedure for constructing a numerical solution which is based on approximating the solution curve by a series of straight lines.

6.3 The Crude Euler Method

An approximate solution to Eq. (6.3) with the initial condition that, at $x = x_0$, $y = y_0$ is illustrated in Fig. 6.2. The true solution is assumed to be known and is shown for comparison.

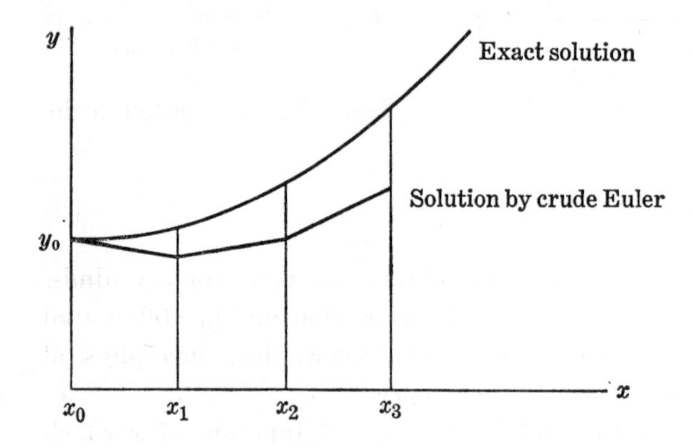

Figure 6.2 Crude Euler solution.

At the point (x_0,y_0) the value for the slope of the solution curve is computed by using the differential equation in (6.3).

$$y'(x_0) = f(x_0,y_0)$$

The solution curve on the range x_0 to x_1 is approximated by the straight line which passes through (x_0,y_0) with the slope $f(x_0,y_0)$. If the value for y at x_1 is designated y_1, then y_1 is given by

$$y_1 = y_0 + (x_1 - x_0)f(x_0,y_0) \tag{6.6a}$$

The point (x_1,y_1) is an approximation to a point on the solution curve. The solution on the interval x_1 to x_2 is approximated by a straight line through (x_1,y_1) with a slope of $f(x_1,y_1)$.

$$y_2 = y_1 + (x_2 - x_1)f(x_1,y_1) \tag{6.6b}$$

The value for y at $x = x_{i+1}$ is obtained from the approximation at x_i.

$$y_{i+1} = y_i + (x_{i+1} - x_i)f(x_i,y_i) \tag{6.6c}$$

The solution constructed in this way is a series of straight lines such as those shown in Fig. 6.2.

This method, called the *crude Euler method*, is the simplest of the numerical techniques and is subject to considerable error compared with some of the more refined methods discussed later. It is used in the following example to illustrate some of the general features of numerical methods for initial-value problems.

Example 6.1

Find the approximate solution to the initial-value problem in Eq. (6.7). Use a spacing h of 0.1.

$$\begin{aligned} y' &= 2xy \qquad 0 \le x \le 1.0 \\ y(0) &= 1 \end{aligned} \tag{6.7}$$

The results from Eqs. (6.6a) to (6.6c) are tabulated in Table 6.1 under y (computed). The slope of the straight line used to approximate the solution on the interval i to $i + 1$ is given in the ith row under y'. The true solution and the error are given for comparison.

The crude Euler method for solving the initial-value problem develops large errors unless the straight-line segments are good approximations to the true curve. If the curvature of the true solution is large, or if the interval from x_i to x_{i+1} is large, this approximation can quickly diverge from the true solution. Considerable

Table 6.1
Solution of y′ = 2 xy by the Crude Euler Method

i	x	y (computed)	y'	y (true)	Error
0	0.0	1.000000	0.000000	1.000000	0.000000
1	0.1	1.000000	0.200000	1.010050	0.010050
2	0.2	1.020000	0.408000	1.040811	0.020811
3	0.3	1.060800	0.636480	1.094174	0.033374
4	0.4	1.124448	0.899558	1.173511	0.049063
5	0.5	1.214404	1.214404	1.284025	0.069621
...	...	...	...	...	...
10	1.0	2.334632	4.669264	2.718282	0.383650

improvement in the approximation can be obtained by a correction of the crude Euler result.

6.4 The Corrected Euler Method

The crude Euler approximation of the previous section is based upon a sequence of straight-line approximations to the solution curve. Each straight line extends the solution by passing through the last point that is computed with a slope equal to the value of $f(x,y)$ at that point. For example, in Fig. 6.2 the approximate value for y_1 is found by passing a straight line through the point (x_0,y_0) with a slope equal to $f(x_0,y_0)$. The error in this procedure can grow rapidly, as is seen from Example 6.1.

An improvement in the crude Euler approximation for any given spacing on x can be made by using a better straight-line approximation to the solution curve or a higher-degree polynomial that better approximates the solution curve in the interval of interest. Both these possibilities are illustrated for the interval x_0 to x_1 in Fig. 6.3. In the corrected Euler method a straight line that approximates the solution curve more closely is used to find the value of y_1. The corrected Euler approximation shown in Fig. 6.3 passes through the point at (x_0,y_0) with a slope that is equal to the average of $f(x_0,y_0)$ and $f(x_1,y_1)$. The value of y_1 used to compute $f(x_1,y_1)$ is obtained from the crude Euler approximation. To find the value for y_1 at x_1 by the corrected Euler method a value for y_1 is first predicted by using Eq. (6.6*a*); call this value p_1. A corrected value for y_1 is then obtained by using the predicted value for y_1 to evaluate $f(x_1,y_1)$ in

$$y_1 = y_0 + (x_1 - x_0)\tfrac{1}{2}[f(x_0,y_0) + f(x_1,p_1)] \tag{6.8}$$

A first approximation to the value for y at x_2, called the predicted value p_2, is made by using the crude Euler approximation. The

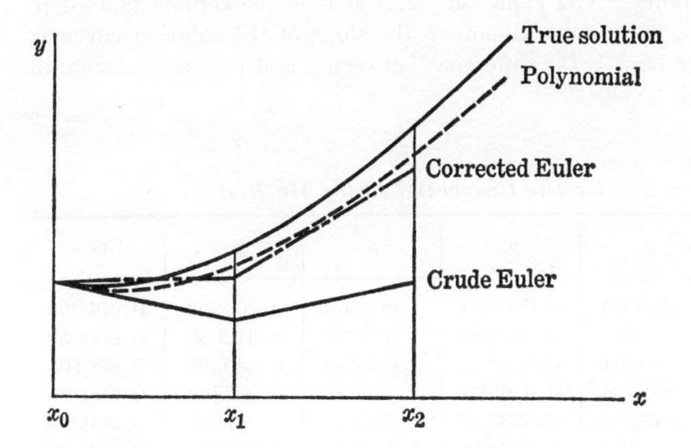

Figure 6.3 Approximations to a solution curve.

predicted value is then used in Eq. (6.9) to obtain a corrected value for y_2.

$$y_2 = y_1 + (x_2 - x_1)\tfrac{1}{2}[f(x_1,y_1) + f(x_2,p_2)] \tag{6.9}$$

The value for y at $x = x_{i+1}$ is found by first predicting a value for y_{i+1}, called p, in Eq. (6.10), and then using the predicted value to obtain a corrected value for y_{i+1}. The subscript on p can be dropped, since it is a temporary value for the solution used only to obtain an improved approximation.

$$\begin{aligned} p &= y_i + hf(x_i,y_i) \\ y_{i+1} &= y_i + \tfrac{1}{2}h[f(x_{i+1},p) + f(x_i,y_i)] \end{aligned} \tag{6.10}$$

where $h = x_{i+1} - x_i$. The procedure of predicting a result by one formula and then correcting by another is one of the most effective methods for solving initial-value problems and is called a *predictor-corrector method.*

Example 6.2

The initial-value problem in Example 6.1 is solved by the formulas of Eqs. (6.10). The results are tabulated in Table 6.2 to illustrate the improvement that is obtained by using the corrected Euler method. These results are much closer to the true solution than the crude Euler approximation. In Table 6.2 the value for p in row i is the estimate of the solution at x_{i+1} based on the crude Euler formula. The value for $f(x_{i+1},p)$ is in the column labeled p' and row i. It is an estimate to the slope of the solution curve at x_{i+1}. The error is the difference between y and the true solution in Table 6.1.

Table 6.2
Solution of $y' = 2xy$ by the Corrected Euler Method

i	x	y	y'	p	p'	Error
0	0.0	1.000000	0.000000	1.000000	0.200000	0.000000
1	0.1	1.010000	0.202000	1.030200	0.412080	0.000050
2	0.2	1.040704	0.416282	1.082332	0.649399	0.000107
3	0.3	1.093988	0.656393	1.159627	0.927702	0.000186
4	0.4	1.173193	0.938554	1.267048	1.267048	0.000318
5	0.5	1.283473	1.283473	1.411820	1.694184	0.000552
...	...	...	...	...	...	...
10	1.0	2.709058	5.418116			0.009224

Further improvement in the numerical results of the method described above is sometimes attempted by using the second of Eqs. (6.10) as an iteration equation for y_{i+1}. The iterations are

Table 6.3
Solution of $y' = 2xy$ by the Iterated Euler Method

i	x	y	y'	Number of iterations	Error
0	0.0	1.000000	0.000000		0.000000
1	0.1	1.010101	0.202020	2	−0.000051
2	0.2	1.041023	0.416409	4	−0.000212
3	0.3	1.094684	0.656810	3	−0.000510
4	0.4	1.174504	0.939603	3	−0.000993
5	0.5	1.284908	1.284908	4	−0.000883
...	...	...	...	...	...
10	1.0	2.727729	5.329008	4	−0.009447

repeated until the difference in the value for y_{i+1} from two successive iterations is smaller than some given number. This method is called the *iterated Euler method.* It is illustrated in Example 6.3, and some results are given in Table 6.3.

Example 6.3

The initial-value problem in Example 6.1 is solved by the iterated Euler method. The iterations are shown here for the solution at $x = 0.1$:

$$p_0 = y(0) + hy'(0) = 1.000000 + 0 = 1.000000$$

$$p_0' = 2xp_0 = 2(0.1)(1.000000) = .200000$$

$$p_1 = y(0) + \tfrac{1}{2}h[y'(0) + p_0'] = 1.000000 + 0.05(0 + 0.200000)$$

$$= 1.010000$$

$$p_1' = 2xp_1 = 0.202000$$

$$p_2 = y(0) + \tfrac{1}{2}h[y'(0) + p_1'] = 1.010100$$

$$p_2' = 2xp_1 = 0.202020$$

$$p_3 = y(0) + \tfrac{1}{2}h[y'(0) + p_2'] = 1.010101$$

Further iterations would not change the sixth decimal in p_3, so p_3 is taken as the value for $y(0.1)$. The results of this method are tabulated in Table 6.3. The number of iterations for each result is given in column five of the table. The error gives the difference between this approximation and the exact solution. It is interesting to note that the iterations do not give any improvement in the solution to this problem. Instead they converge to values which are poorer approximations than the corrected Euler approximation used to start the iteration (see Table 6.2). In general, iterations with the corrector formula should be used with caution.

A flow chart for finding m points in the solution of Example 6.1 using the corrected Euler method is given in Fig. 6.4.

The crude, corrected, and iterated Euler methods have been developed by intuitive arguments which are based on the nature of the initial-value problem. There is no provision in the development for improving the approximation or for finding the error associated

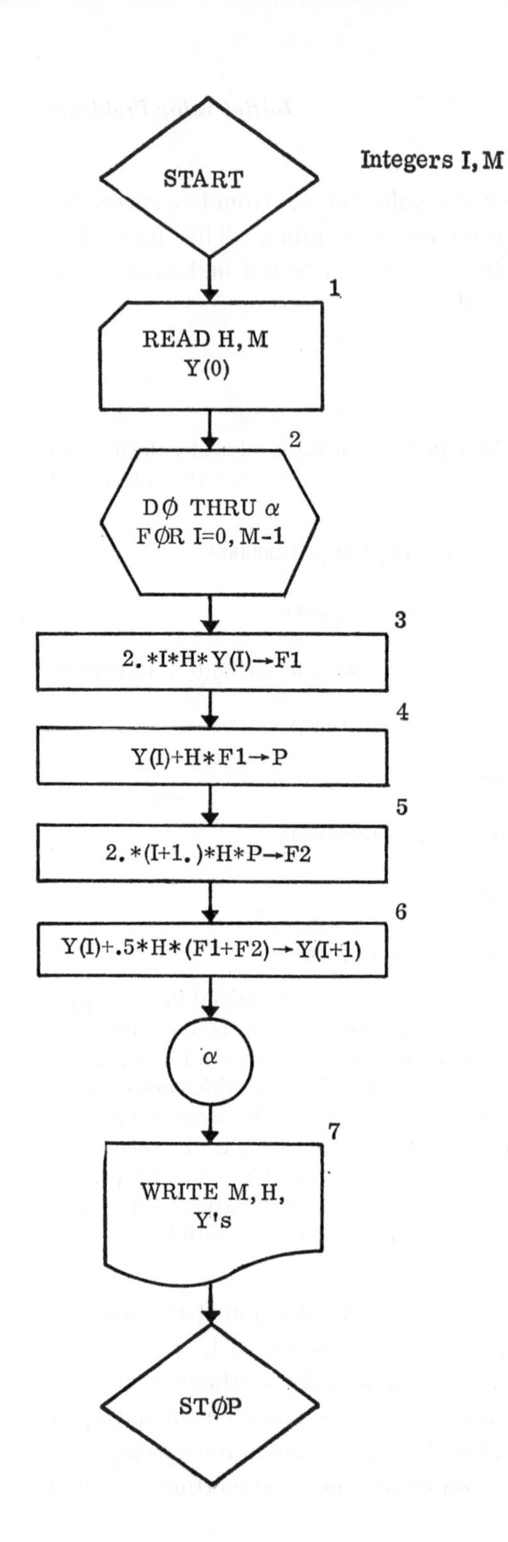

Figure 6.4 Corrected Euler method for $y' = 2xy$, $y(0) = 1$.

with the approximation. There exist formal methods for finding these formulas as well as a large family of other formulas of the predictor-corrector type. A direct result of the formal methods for constructing the solution formulas is an expression for the truncation error of the approximation. In the next sections formal procedures for approximating the solution of the initial-value problem are developed.

6.5 *Predictor Formulas*

The general predictor formula for solving differential equations is based on the fundamental theorem of the integral calculus stated in

$$\int_{x_0}^{x_1} y'(x)\,dx = y(x_1) - y(x_0) \tag{6.11}$$

If $y(x_1)$ is to be the solution of the initial-value problem (6.3), then $y(x_1)$ must satisfy Eq. (6.12). Let $y_0 = y(x_0)$ and $y_1 = y(x_1)$; then

$$y_1 = y_0 + \int_{x_0}^{x_1} f(x,y(x))\,dx \tag{6.12}$$

y_1 is the solution of the differential equation (6.3) at $x = x_1$ under the condition that at $x = x_0$ the value for y is y_0. Equation (6.12) will give the value for y at $x = x_{i+1}$ in terms of the value for y at $x = x_{i-r}$ if x_0 is replaced by x_{i-r} and x_1 by x_{i+1}.

$$y_{i+1} = y_{i-r} + \int_{x_{i-r}}^{x_{i+1}} f(x,y)\,dx \tag{6.13}$$

Equation (6.13) is the basis upon which the formal derivation for the numerical solution of the initial-value problem is made. If the value for y is known at $x = x_{i-r}$, then y_{i+1} is obtained by evaluating the integral in Eq. (6.13). Approximations to the integral are made by using the integration formulas in Chap. 5. In the approximation to the integral it is assumed that values for $y_i, y_{i-1}, \ldots, y_0$ have already been found. The problem is then one of extending the solution to the next point, i.e., finding y_{i+1}.

The approximation to the integral in (6.13) is obtained from the general integration formula in Eq. (5.41). The upper limit on α

is set equal to 1, the lower limit to r, and the following approximation in terms of backward differences is obtained:

$$\begin{aligned}\int_{x_{i-r}}^{x_{i+1}} f(x,y)\,dx = h[(1+r)f_i + \tfrac{1}{2}(1-r^2)\,\Delta f_i \\ + \tfrac{1}{12}(5-3r^2+2r^3)\,\Delta^2 f_i \\ + \tfrac{1}{24}(9-4r^2+4r^3-r^4)\,\Delta^3 f_i \\ + \tfrac{1}{720}(251-90r^2+110r^3-45r^4+6r^5)\,\Delta^4 f_i + \cdots]\end{aligned} \qquad (6.14)$$

where $f_i = f(x_i,y_i)$. The truncation error when $m+1$ terms are used in the approximation is

$$E_m(r) = \frac{h^{m+2}}{(m+1)!}\int_{-r}^{1} \alpha(\alpha+1)\cdots(\alpha+m)y^{(m+2)}(\xi)\,d\alpha \qquad (6.15)$$

where $x_{i-r} \le \xi \le x_{i+1}$. The formula in Eq. (6.14) is an open-end integration formula, since the upper limit of integration extends beyond the known points used to construct the backward-difference approximation to the integrand. It should be noted that Eqs. (6.14) and (6.15) are applicable only when the spacing on x is constant and equal to h.

Many approximate formulas for y_{i+1} can be obtained by using different values for r and varying the number of terms used in Eq. (6.14). For example, the crude Euler formula is obtained by setting $r = 0$ and using only one term in the approximation. The geometry of the approximation is illustrated in Fig. 6.2.

$$y_{i+1} = y_i + hf_i \qquad (6.6)$$

The truncation error in this formula is obtained by letting $m = 0$ and $r = 0$ in Eq. (6.15).

$$E_0(0) = \tfrac{1}{2}h^2y''(\xi) \qquad (6.16)$$

Thus the crude Euler formula has an $O(h^2)$ truncation error, which means that the error in the numerical solution at any given point is $O(h)$. Thus if the spacing is divided by 2, the truncation error in the solution at a given point will also be divided by 2 (see Sec. 5.7).

An open-end formula with a truncation error $O(h^3)$ is obtained by taking two terms of Eq. (6.14) and setting $r = 1$.

$$\begin{aligned} y_{i+1} &= y_{i-1} + 2hf_i \\ E_1(1) &= \tfrac{1}{3}h^3y'''(\xi) \end{aligned} \qquad (6.17)$$

This is the predictor formula that is used for the Euler predictor-corrector method. It gives a value for y_{i+1} in terms of the solution at x_i and x_{i-1} with a truncation error of $O(h^3)$.

Figure 6.5 shows geometrically how the solution is extended by Eqs. (6.17). The solution curve is approximated on the interval x_{i-1} to x_{i+1} by the straight line having a slope equal to $f(x_i,y_i)$. In order to apply this formula two consecutive values for the solution must be known. For example, to obtain the solution y_1 at $x = x_0 + h$ the value for y at $x = x_0$ and $x = x_0 - h$ must be known. One of the starting values is the initial condition, but the other must be computed by some other numerical procedure. This

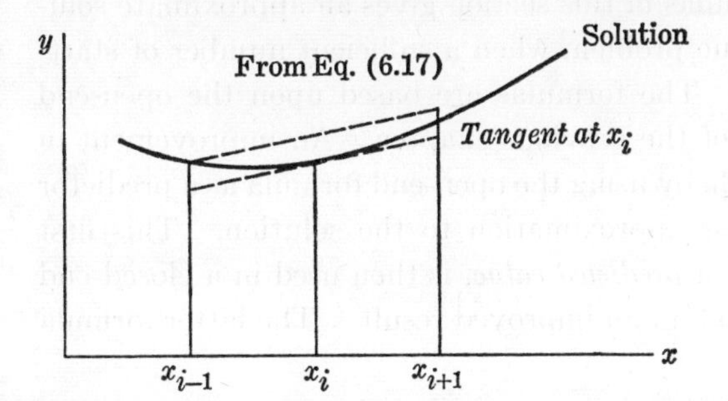

Figure 6.5 A predictor with $O(h^3)$ truncation error.

is a characteristic of the higher-order methods and is a major consideration in their use. At this point in the development of the methods it is assumed that some suitable technique will be available for computing as many starting values as are required. For application of Eqs. (6.17) it is sufficient to assume that y_0 and y_1 are known.

Higher-order approximations to y_{i+1} are readily obtained from Eq. (6.14) by using more terms. A predictor formula used in the Adams-Bashforth predictor-corrector method (Sec. 6.9) is obtained by using four terms in the approximation in Eq. (6.14) and setting $r = 0$.

$$y_{i+1} = y_i + \tfrac{1}{24}h[55f_i - 59f_{i-1} + 37f_{i-2} - 9f_{i-3}] + E_4(0) \tag{6.18}$$

$$E_4(0) = \tfrac{251}{720}h^5y^{V}(\xi)$$

The application of Eqs. (6.18) requires four starting values. The solution at $x = x_{i+1}$ requires the solution at the four points $x = x_i$, x_{i-1}, x_{i-2}, and x_{i-3}. This approximation to the solution is an improvement over the linear estimates used in the preceding formulas and is reflected in the higher-order truncation error.

Many approximate formulas can be obtained by varying r, which controls the interval of integration, and by varying the number of terms in the approximation, which controls the order of the approximation. In general to obtain an $O(h^n)$ truncation error requires that $n - 1$ terms be used in the formula for the integral.

Each of the formulas in this section gives an approximate solution to the initial-value problem when a sufficient number of starting values is known. The formulas are based upon the open-end integration formulas of the previous chapter. An improvement in the results can be made by using the open-end formula as a predictor formula to give a first approximation to the solution. This first approximation, called a *predicted value*, is then used in a closed-end integration formula to find an improved result. The latter formula is called a *corrector*.

6.6 Corrector Formulas

The corrector formulas are based on using a closed-end integration formula to evaluate the integral in Eq. (6.13). In order to obtain a closed-end integration formula for the integral the backward-difference approximation for $f(x,y)$ at $x = x_{i+1}$ is used for the integrand.

Figure 6.6 shows a plot of the function $f(x,y)$. The function is approximated by the backward-difference formula through m points at x_{i+1-m}, x_{i+2-m}, . . . , x_{i+1}. The solution y is assumed to be known at each value of x up to and including $x = x_i$. An estimate of the solution at $x = x_{i+1}$ is obtained from one of the predictor formulas in Sec. 6.5. The corrector formula for y_{i+1} is then obtained from Eq. (6.13) by making the integration from x_{i+1-r} to x_{i+1}.

$$y_{i+1} = y_{i+1-r} + \int_{x_{i+1-r}}^{x_{i+1}} f(x,y)\, dx \tag{6.19}$$

The backward-difference formula in terms of differences at $x = x_{i+1}$ is used to replace $f(x,y)$ in Eq. (6.19). The limits of integration on α are $\alpha = -r$ to $\alpha = 0$.

$$y_{i+1} = y_{i+1-r} + h \int_{-r}^{0} \left[f_{i+1} + \alpha\, \underline{\Delta} f_{i+1} + \frac{\alpha(\alpha+1)}{2!}\, \underline{\Delta}^2 f_{i+1} + \frac{\alpha(\alpha+1)(\alpha+2)}{3!}\, \underline{\Delta}^3 f_{i+1} + \cdots + \frac{h^{m+1}}{(m+1)!}\, \alpha(\alpha+1) \cdots (\alpha+m) f^{(m+1)}(\xi)\right] d\alpha \qquad (6.20)$$

where

$$x = x_{i+1} + \alpha h \qquad \text{and} \qquad f_{i+1} = f(x_{i+1}, y_{i+1})$$

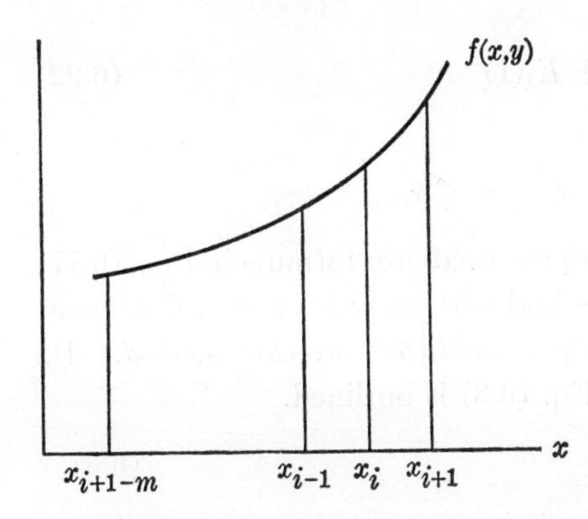

Figure 6.6 Backward-difference approximation to $f(x,y)$.

The basic equation from which the corrector formulas can be derived is obtained by integrating Eq. (6.20). The result is

$$y_{i+1} = y_{i+1-r} + h\left[r f_{i+1} - \frac{r^2}{2}\, \underline{\Delta} f_{i+1} - \frac{r^2}{12}(3 - 2r)\, \underline{\Delta}^2 f_{i+1} - \frac{r^2}{24}(2 - r)^2\, \underline{\Delta}^3 f_{i+1} - \frac{r^2}{720}(-6r^3 + 45r^2 - 110r + 90)\, \underline{\Delta}^4 f_{i+1} + \cdots\right] + E_m(r) \qquad (6.21)$$

where

$$E_m(r) = \frac{h^{m+2}}{(m+1)!} \int_{-r}^{0} \alpha(\alpha+1)(\alpha+2) \cdots (\alpha+m) f^{(m+1)}(\xi)\, d\xi$$

By varying r and the number of terms in the approximation a large number of formulas can be obtained. The principal objective here is to derive the corrector formulas that will pair with the predictor formulas of the previous section so that they can be used together in a predictor-corrector method. The predictor and corrector formulas in a predictor-corrector method should have truncation errors of the same order, and the formulas should be kept as simple as possible.

6.7 Euler Predictor-Corrector Method

A corrector formula with an $O(h^3)$ error is obtained from Eq. (6.21) by setting $r = 1$ and using two terms in the approximation.

$$y_{i+1} = y_i + \frac{h}{2}(f_i + f_{i+1}) + E_1(1) \tag{6.22}$$

where

$$E_1(1) = -\tfrac{1}{12}h^3y'''(\xi) \qquad x_i \leq \xi \leq x_{i+1}$$

Equation (6.22) is combined with the predictor formula in Eqs. (6.17) to give a predictor-corrector method with an $O(h^3)$ truncation error. This method is called the *Euler predictor-corrector method.* Its application to the problem in Eq. (6.3) is outlined.

$$y' = f(x,y) \qquad y(x_0) = y_0 \tag{6.3}$$

For purposes of illustration it is assumed that the solution at $x = x_1$ is known and is equal to y_1. Equations (6.17) are used to find an estimate to y_2, which is designated p.

$$p = y_0 + 2hf(x_1,y_1) \tag{6.23}$$

The estimate to y_2 in Eq. (6.23) is used in Eq. (6.22) to give the value for y_2.

$$y_2 = y_1 + \frac{h}{2}[f(x_1,y_1) + f(x_2,p)] \tag{6.24}$$

The solution at $x = x_{i+1}$, for $i = 1, 2, 3, \ldots$, is given by

$$p = y_{i-1} + 2hf(x_i,y_i) \tag{6.25a}$$

$$y_{i+1} = y_i + \frac{h}{2}[f(x_i,y_i) + f(x_{i+1},p)] \tag{6.25b}$$

Example 6.4

The initial-value problem in Example 6.1 is solved by the Euler predictor-corrector method. The value for y_1 is obtained from the corrected Euler method, and the results are tabulated in Table 6.4. The column labeled y' gives the value for $f(x,y)$ at the tabular values for i. The predicted value for the solution at $i + 1$ that is obtained from Eq. (6.25a) is tabulated in the ith row under p. Thus the predicted value of the solution at $x = 0.3$ is given in row $i = 2$. A comparison of the errors in Table 6.4 with the errors in the previous methods (Tables 6.1 to 6.3) shows that this method has a larger error than some of the others. This is not unexpected, since it can be

Table 6.4
Solution of $y' = 2xy$ by the Euler Predictor-Corrector Method

i	x	y	y'	p	Error
0	0.0	1.000000	0.000000		0.000000
1	0.1	1.010000	0.202000	1.040400	0.000050
2	0.2	1.040908	0.416363	1.093273	−0.000097
3	0.3	1.094523	0.656714	1.172251	−0.000349
4	0.4	1.174249	0.939399	1.282403	−0.000738
5	0.5	1.285339	1.285339	1.431317	−0.001314
. . .	. . .	. . .	. . .	. . .	. . .
10	1.0	2.730196	5.460392		−0.011914

shown that the corrected Euler method also has a truncation error of $O(h^3)$. The accuracy of the Euler predictor-corrector method can be improved significantly by a modification which is given in the next section.

6.8 *Modified Euler Predictor-Corrector Method*

A considerable improvement can be made in the numerical results of the Euler predictor-corrector method by making a crude estimate of the truncation error in the corrector formula and then adding the correction to the final result. This method is called the *modified Euler predictor-corrector method.* In order to find an estimate to the error it is assumed that the derivative in the error terms of Eqs. (6.17) and (6.22) is slowly varying on the interval of approximation and can be replaced by a constant.

The solution at y_{i+1} is equal to the predictor value p plus the truncation error in the formula

$$y_{i+1} = p + \frac{h^3}{3} y'''(\xi) \tag{6.26a}$$

The value for y_{i+1} is also equal to the approximation from Eqs. (6.25*a*) and (6.25*b*), designated c, plus the error in the corrector formula.

$$y_{i+1} = c - \frac{h^3}{12} y'''(\xi) \tag{6.26b}$$

Equations (6.26*a*) and (6.26*b*) are solved for $y'''(\xi)$. The result is substituted into Eq. (6.26*b*) to give the final corrected value for y_{i+1}.

$$y_{i+1} = \tfrac{1}{5}(4c + p) \tag{6.27}$$

In the modified predictor-corrector method a predicted value p for y_{i+1} is computed by using Eq. (6.25*a*). Then a corrected value c is computed by using Eq. (6.25*b*). Finally the value for y_{i+1} is found by using Eq. (6.27). The three equations for the modified Euler predictor-corrector method are collected as follows:

$$p = y_{i-1} + 2hf(x_i,y_i) \tag{6.28a}$$

$$c = y_i + \tfrac{1}{2}h[f(x_i,y_i) + f(x_{i+1},p)] \tag{6.28b}$$

$$y_{i+1} = \frac{4c + p}{5} \tag{6.28c}$$

Example 6.5

The initial-value problem of Example 6.1 is solved using the modified Euler predictor-corrector method. The value for y_1 is obtained using the corrected Euler method. The results are given in Table 6.5.

The error columns from Tables 6.1 to 6.5 are presented in Table 6.6 for direct comparison. Clearly, the modified Euler predictor-corrector method gives the best results.

On the basis of experience with numerous problems the modified Euler predictor-corrector method appears to give better results than the other third-order predictor-corrector methods. It has also been

Table 6.5
Solution of $y' = 2xy$ by the Modified Euler Predictor-Corrector Method

i	x	y	y'	p	c	Error
0	0.0	1.000000	0.000000			0.000000
1	0.1	1.010000	0.212000	1.040400	1.040908	0.000050
2	0.2	1.040806	0.416322	1.093265	1.094420	0.000005
3	0.3	1.094189	0.656513	1.172109	1.173899	−0.000015
4	0.4	1.173541	0.938833	1.281956	1.284581	−0.000030
5	0.5	1.284056	1.284056	1.430355	1.434080	−0.000031
. . .	. . .	. . .	. . .	. . .	. . .	. . .
10	1.0	2.717411	5.434822			0.000871

Table 6.6
Errors in the Euler Methods

i	x	Crude	Corrected	Iterated	Predictor-corrector	Modified predictor-corrector
0	0.0	0.000000	0.000000	0.000000	0.000000	0.000000
1	0.1	0.010050	0.000050	−0.000051	0.000050	0.000050
2	0.2	0.020811	0.000107	−0.000212	−0.000097	0.000005
3	0.3	0.033374	0.000186	−0.000510	−0.000349	−0.000015
4	0.4	0.049063	0.000318	−0.000993	−0.000738	−0.000030
5	0.5	0.069621	0.000552	−0.000883	−0.001314	−0.000031
. . .	. . .	. . .	. . .	. . .	. . .	. . .
10	1.0	0.383650	0.009224	−0.009447	−0.011914	0.000871

found that the propagation of an error from one iteration to the next can be reduced by using the modified method.

A flow chart to solve the initial-value problem in Eqs. (6.29) by the modified Euler predictor-corrector method is given in Fig. 6.7. In this flow chart the starting value needed for the Euler predictor-corrector method is obtained by using the corrected Euler method in Eqs. (6.10).

$$\begin{aligned} y' &= -\tfrac{1}{4}[x + (x^2 + 8y)^{1/2}] \\ y(x_0) &= y_0 \end{aligned} \qquad (6.29)$$

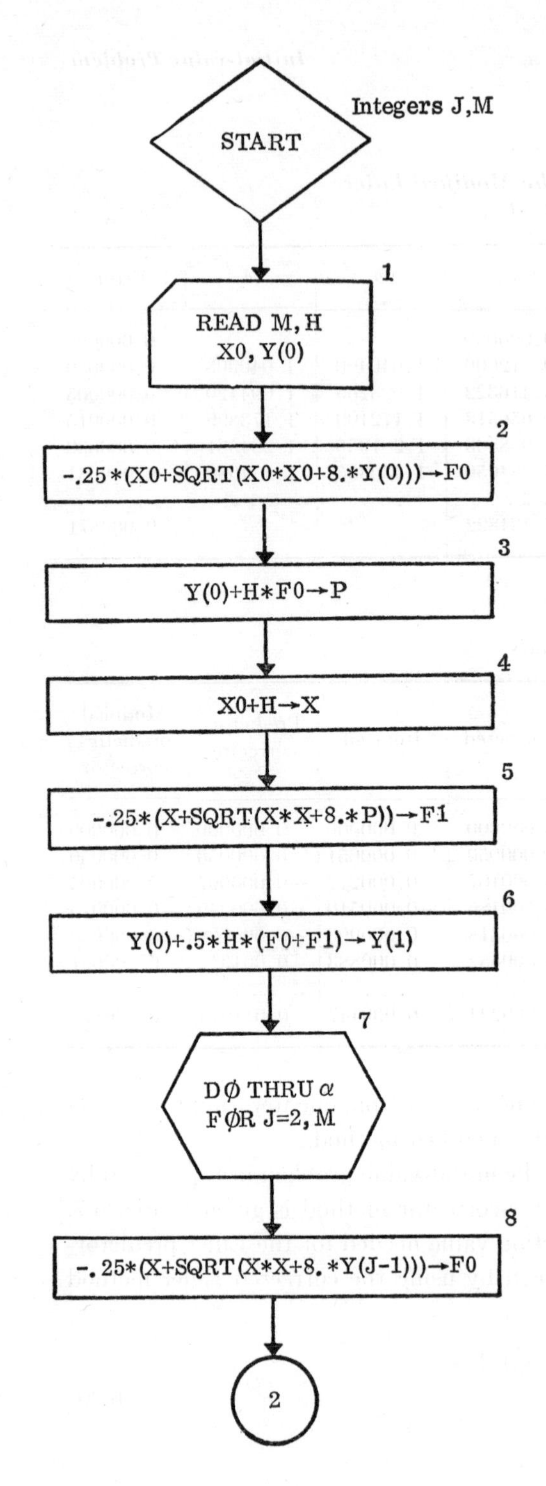

Figure 6.7 Modified Euler predictor-corrector method.

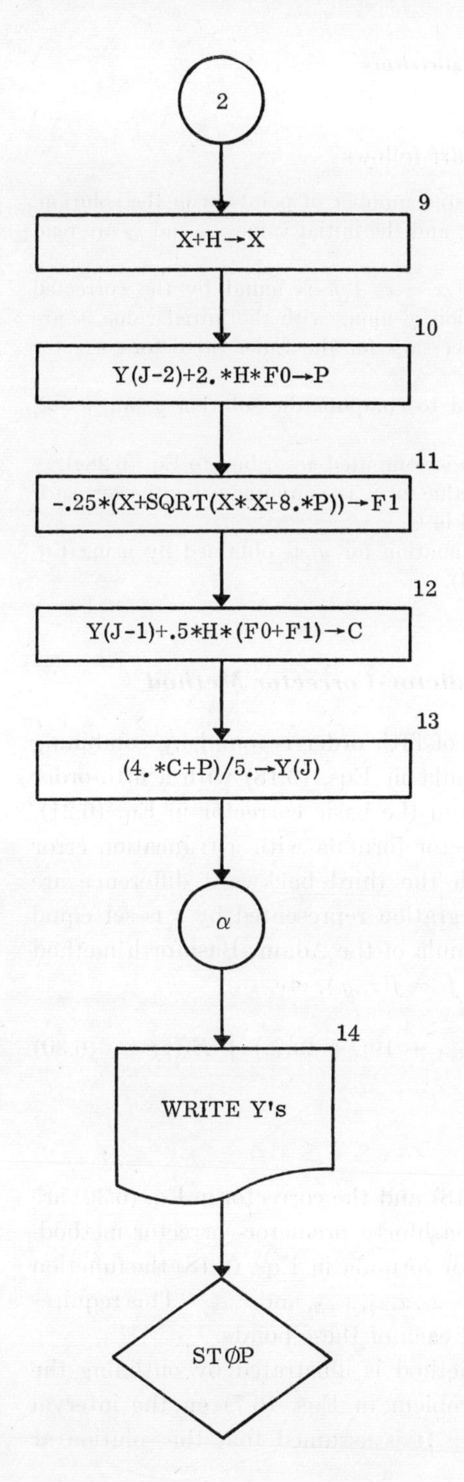

Figure 6.7 *(Continued)*

An explanation of the flow chart follows

1. Values for the desired number of points m in the solution, the spacing h on the x axis, and the initial values x_0 and y_0 are read into the computer.

2–6. The solution at $x = x_0 + h$ is found by the corrected Euler method. This solution y_1 along with the initial value y_0 are the two starting values necessary for the Euler predictor-corrector method.

7. This is the control to compute the solution y_j for $j = 2, \ldots, m$.

8, 10. The value of p is computed according to Eq. (6.28*a*).

11, 12. A corrected value for y_j is computed using the corrector in Eq. (6.28*b*) and is stored in C.

13. The final approximation for y_j is obtained by using Eq. (6.28*c*) and is stored in Y(J).

6.9 Adams-Bashforth Predictor-Corrector Method

A predictor-corrector method of fifth order is found by combining the fifth-order predictor formula in Eqs. (6.18) with a fifth-order corrector formula obtained from the basic corrector in Eq. (6.21). Equation (6.21) gives a corrector formula with a truncation error of $O(h^5)$ when terms through the third backward difference are retained. If the limit of integration represented by r is set equal to 1, the special corrector formula of the Adams-Bashforth method is obtained. Here, as before, $f_i = f(x_i,y_i)$, etc.

$$y_{i+1} = y_i + \frac{h}{24}(f_{i-2} - 5f_{i-1} + 19f_i + 9f_{i+1}) + E_3(1) \qquad (6.30)$$

where

$$E_3(1) = -{}^{19}\!/_{720}h^5y^V(\xi) \qquad x_{i-2} \le \xi \le x_{i+1}$$

The predictor in Eqs. (6.18) and the corrector in Eq. (6.30) are the formulas for the Adams-Bashforth predictor-corrector method. In order to apply the predictor formula in Eqs. (6.18) the function $f(x,y)$ must be evaluated at $x = x_i, x_{i-1}, x_{i-2}$, and x_{i-3}. This requires that the solution be known at each of these points.

The Adams-Bashforth method is illustrated by outlining the procedure for solving the problem in Eqs. (6.7) on the interval $0 \le x \le b$, with a spacing h. It is assumed that the solution at

$x = x_1, x_2, x_3$ has been computed by some other means and will serve as starting values.

$$y' = 2xy \qquad 0 \le x \le b$$
$$y(0) = y_0 = 1 \tag{6.7}$$

For convenient reference the predictor and corrector are repeated here:

$$p = y_i + \frac{h}{24}[55f_i - 59f_{i-1} + 37f_{i-2} - 9f_{i-3}]$$
$$y_{i+1} = y_i + \frac{h}{24}(f_{i-2} - 5f_{i-1} + 19f_i + 9f_{i+1}) \tag{6.30a}$$

Here $f_k = f(x_k,y_k)$, $k = i - 3, i - 2, i - 1, i$, and $f_{i+1} = f(x_{i+1},p)$. The calculation sequence for the solution of Eqs. (6.7) is outlined below.

1. Compute the numerical value of f_0, f_1, f_2, and f_3 using the given starting values:

$$f_0 = 0$$
$$f_1 = 2hy_1$$
$$f_2 = 4hy_2$$
$$f_3 = 6hy_3$$

2. Find the predicted value for the solution y_4:

$$p = y_3 + \frac{h}{24}(55f_3 - 59f_2 + 37f_1 - 9f_0)$$

3. Use the predicted value for y_4 to find an approximate value for f_4:

$$f_4 = 8hp$$

4. Compute y_4 with the corrector formula:

$$y_4 = y_3 + \frac{h}{24}(f_1 - 5f_2 + 19f_3 + 9f_4)$$

5. Find the corrected value for f_4:

$$f_4 = 8hy_4$$

6. Repeat steps 2 to 5 with updated subscripts to find y_5.

By making m repetitions of steps 2 to 5 and updating the subscripts on each iteration, m points in the solution are generated.

6.10 Modified Adams-Bashforth Predictor-Corrector Method

An estimate of the truncation error in the corrector formula for the Adams-Bashford method can be made in a way that is similar to the Euler predictor-corrector method in Sec. 6.8. If the error is added to the corrector in Eq. (6.30), an improved approximation to the solution is obtained.

Equations (6.18) and (6.30) give the solution at $x = x_{i+1}$ in terms of an estimate to y_{i+1} and an error term. The estimate in (6.18) is the predicted value and is designated p; whereas Eq. (6.30) gives a corrected value which will be designated c. The true value for y_{i+1} is the approximation plus the error and is given by both equations as follows:

$$\begin{aligned} y_{i+1} &= p + {}^{251}\!/_{720} h^5 y^{\mathrm{v}}(\xi_1) \\ y_{i+1} &= c - {}^{19}\!/_{720} h^5 y^{\mathrm{v}}(\xi_2) \end{aligned} \tag{6.31}$$

It is assumed that the fifth derivative of the solution is slowly varying on the interval and can be approximated by a constant. On the basis of this assumption the error in the second of Eqs. (6.31) is evaluated in terms of p and c to give

$$y_{i+1} = c + {}^{19}\!/_{270}(p - c) = \frac{251c + 19p}{270} \tag{6.32}$$

The Modified Adams-Bashforth predictor-corrector method uses Eqs. (6.30*a*) to obtain a predicted value p and a corrected value c for y_{i+1}, and then uses Eq. (6.32) to obtain y_{i+1}. These equations are repeated here for convenience.

$$\begin{aligned} p &= y_i + \frac{h}{24}(55f_i - 59f_{i-1} + 37f_{i-2} - 9f_{i-3}) \\ c &= y_i + \frac{h}{24}[9f(x_{i+1},p) + 19f_i - 5f_{i-1} + f_{i-2}] \\ y_{i+1} &= \frac{251c + 19p}{270} \end{aligned} \tag{6.33}$$

where

$$f_k = f(x_k, y_k) \qquad \text{for } k = i-3, i-2, i-1, i$$

Example 6.6

The initial-value problem in Example 6.1 is solved using the modified Adams-Bashforth predictor-corrector method.

$$y' = 2xy \qquad 0 \le x \le 1$$
$$y(0) = 1 \tag{6.7}$$

The solution is obtained with $h = 0.1$. The necessary starting values are obtained by the Runge-Kutta method (Sec. 6.12), and the results are summarized in Table 6.7.

Table 6.7
Solution of $y' = 2xy$ by the Modified Adams-Bashforth Predictor-Corrector Method

i	x	y	p	c	Error
0	0.0	1.0000000			0.0000000
1	0.1	1.0100502*			0.0000000
2	0.2	1.0408108*			0.0000000
3	0.3	1.0941743*	1.1734200	1.1735184	0.0000000
4	0.4	1.1735115	1.2838725	1.2840366	−0.0000006
5	0.5	1.2840250	1.4330891	1.4333429	0.0000004
6	0.6	1.4333250	1.6319474	1.6323293	0.0000044
7	0.7	1.6323024	1.8959170	1.8964879	0.0000138
8	0.8	1.8964477	2.2470419	2.2478975	0.0000332
9	0.9	2.2478372	2.7169396	2.7182313	0.0000708
10	1.0	2.7181403			0.0001415

* Obtained by the Runge-Kutta Method.

A comparison of the results in Table 6.7 with previous methods shows that the Adams-Bashforth method gives nearly an order of magnitude smaller error than the best of the third-order methods.

A flow chart for the modified Adams-Bashforth predictor-corrector method is given in Fig. 6.10. The starting values for the solution are obtained from the fourth-order Runge-Kutta method (Sec. 6.13).

6.11 *Starting Values for Predictor-Corrector Methods*

The predictor-corrector methods have a drawback in that more than one starting value is necessary. From the initial condition one point in the solution is known. The rest of the starting values must be found from some technique that will give results to the same order of approximation as the predictor-corrector routine that is used. A general approach for finding starting values for the predictor-corrector formulas is based on a class of methods called *Runge-Kutta methods,* which not only provide a means for constructing starting values for the predictor-corrector formulas but are also used extensively to find the complete solution.

In the next section a Runge-Kutta formula with $O(h^3)$ truncation error is developed. Following this the Runge-Kutta formula with $O(h^5)$ truncation error is given.

6.12 *Runge-Kutta Methods*

The Runge-Kutta methods are similar in concept to the Euler methods in that the approximation to the next point in the solution

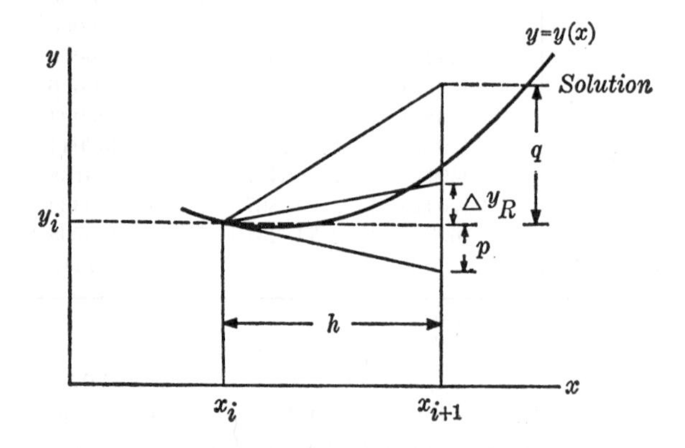

Figure 6.8 A Runge-Kutta method.

is based on a linear approximation to the solution curve. The Runge-Kutta methods use a weighted average for the value of the slope on the range x_i to x_{i+1} to find a value for the change in y. This is illustrated by Δy_R in Fig. 6.8.

A Runge-Kutta method is developed from intuitive arguments and then a formal method of derivation is presented. The intuitive method for finding the value for y_{i+1} when y_i is known is outlined in a step-by-step procedure.

1. The first step is to approximate the change in y by using the crude Euler formula. This is p in Fig. 6.8.

$$p = hf(x_i,y_i) \tag{6.34}$$

2. The value for p is used to approximate the slope of the solution curve at x_{i+1}, i.e., $y'(x_{i+1}) = f(x_i + h, y_i + p)$. The change in y based on the approximation to $f(x,y)$ at x_{i+1} is denoted by q (see Fig. 6.8).

$$q = hf(x_i + h, y_i + p) \tag{6.35}$$

3. The final approximation for y_{i+1} is found by using the average of p and q for the change in y. The change is designated Δy_R in Fig. 6.8.

$$y_{i+1} = y_i + \tfrac{1}{2}(p + q) \tag{6.36}$$

Equation (6.36) is based on a straight-line approximation to the solution curve on x_i to x_{i+1}. The slope of the straight line is the average of the slope at x_i and an approximation to the slope at x_{i+1}. It is shown in the formal development that the formula in Eq. (6.36) has a truncation error of $O(h^3)$. This method is a second-order Runge-Kutta method [the truncation error in the solution is $O(h^2)$] and is exactly equivalent to the corrected Euler method.

A formal derivation of a second-order Runge-Kutta method is now made. The derivation is based on using the Taylor-series expansion for the solution y about the point (x_i,y_i).

$$y(x) = y(x_i + \Delta x) = y(x_i) + \Delta x\, y'(x_i) + \frac{1}{2!}\Delta x^2\, y''(x_i) + \frac{1}{3!}\Delta x^3\, y'''(x_i) + \cdots + \frac{1}{m!}\Delta x^m\, y^m(x_i) + \cdots \tag{6.37}$$

Equation (6.37) will give an approximation to $y(x)$ with a truncation error of $O(\Delta x^m)$ provided m terms are used in the series. An $O(h^m)$ approximation to the solution of the initial-value problem at $x = x_{i+1}$ is obtained in terms of the solution at $x = x_i$ if Δx is set equal to h. Thus,

$$y_{i+1} = y_i + hf(x_i,y_i) + \frac{1}{2!}h^2f'(x_i,y_i) + \frac{1}{3!}h^3f''(x_i,y_i) + \cdots + \frac{1}{m!}h^mf^{(m-1)}(\xi) \quad (6.38)$$

where ξ is on the interval x_i to x_{i+1}. The derivatives of y have been replaced from Eq. (6.3).

If the derivatives (with respect to x) of the function f are readily computed, the Taylor series in Eq. (6.38) can be used directly to find the solution at $x = x_{i+1}$. This computation of the derivative of f is a tedious job at best, and therefore Eq. (6.38) is not in general suitable for finding y_{i+1}. The Runge-Kutta methods are based on constructing formulas which will give the same order of approximation as the Taylor series but are expressed entirely in terms of the function f and not its derivatives.

The second-order Runge-Kutta formula with truncation error of $O(h^3)$ is obtained from

$$y_{i+1} = y_i + a_1p + a_2q \quad (6.39)$$

where

$$p = hf(x_i,y_i)$$

$$q = hf(x_i + \alpha h,\ y_i + \alpha p) \qquad 0 \le \alpha \le 1$$

The constants a_1, a_2, and α are chosen so that the formula in Eq. (6.39) is $O(h^3)$. This is accomplished as follows. Let the function q be expanded in a Taylor series about (x_i,y_i). Then

$$q = h\left[f(x_i,y_i) + \alpha h\frac{\partial f(x_i,y_i)}{\partial x} + \alpha p\frac{\partial f(x_i,y_i)}{\partial y} + O(h^2)\right] \quad (6.40)$$

and

$$y_{i+1} = y_i + a_1hf(x_i,y_i) + a_2h\left[f(x_i,y_i) + \alpha h\frac{\partial f(x_i,y_i)}{\partial x} + \alpha hf(x_i,y_i)\frac{\partial f(x_i,y_i)}{\partial y} + O(h^2)\right] \quad (6.41)$$

In Eq. (6.38) $f'(x_i,y_i)$ is replaced by the equivalent expression

$$f'(x_i,y_i) = \frac{\partial f(x_i,y_i)}{\partial x} + \frac{\partial f(x_i,y_i)}{\partial y}f(x_i,y_i)$$

and is then compared with Eq. (6.41). Each equation gives precisely the same result through terms of $O(h^2)$ if the following equalities are satisfied:

$$\begin{aligned} a_1 + a_2 &= 1 \\ a_2\alpha &= \tfrac{1}{2} \end{aligned} \tag{6.42}$$

Thus when the conditions in Eq. (6.42) are satisfied, the truncation error in Eq. (6.39) is of the same order as the $O(h^3)$ Taylor-series expansion about (x_i,y_i).

Equation (6.39), where the constants a_1, a_2, and α satisfy Eqs. (6.42), gives the Runge-Kutta method of second order. There is some liberty in the choice of the constants for the Runge-Kutta formula. If α is set equal to 1, then $a_1 = a_2 = \tfrac{1}{2}$, and the Runge-Kutta formula that is equivalent to the corrected Euler method of Sec. 6.4 is obtained.

6.13 A Fourth-order Runge-Kutta Method

The development of the fourth-order Runge-Kutta formulas is a tedious algebraic problem and is not undertaken in this text. Like the second-order method in the last section there is some arbitrary choice in the makeup of the formula so that many fourth-order formulas can be developed. One formula, called the *Kutta-Simpson formula,* is comparatively simple and is by far the most popular. This formula is given without proof:

$$\begin{aligned} y_{i+1} &= y_i + \tfrac{1}{6}(p + 2q + 2r + s) + O(h^5) \\ p &= hf(x_i,y_i) \\ q &= hf(x_i + \tfrac{1}{2}h,\ y_i + \tfrac{1}{2}p) \\ r &= hf(x_i + \tfrac{1}{2}h,\ y_i + \tfrac{1}{2}q) \\ s &= hf(x_i + h,\ y_i + r) \end{aligned} \tag{6.43}$$

The values p, q, r, and s are each estimates of the change in y on the interval x_i to x_{i+1}. p and s are estimates based on the slope at the beginning and end of the interval, respectively, while q and r are based on approximations to the slope at the midpoint of the interval. The term that is added to y_i is a weighted average of the several estimates. It is noted that when the function f in Eqs. (6.43) does not contain y, the formula reduces to Simpson's one-third rule for integration, given in Chap. 5.

When using Eqs. (6.43), one must find the terms p, q, r, and s for each point in the solution. This means the function $f(x,y)$ must in general be computed for four different sets of arguments. Should the computation of $f(x,y)$ be involved, it would detract from the usefulness of the method. On the other hand, it is a single-point method which has distinct advantages over the predictor-corrector methods. More will be said about the advantages and disadvantages of the Runge-Kutta methods in Sec. 6.17.

Example 6.7

The initial-value problem in Example 6.1 is solved using the Kutta-Simpson method. The results are tabulated in Table 6.8. The values of p, q, r, and s used to find y_{i+1} are tabulated in row i.

Table 6.8
Solution of $y' = 2xy$ by the Kutta-Simpson Method

i	x	y	p	q	r	s	Error
0	0.0	1.0000000	0.0000000	0.0100000	0.0100500	0.0202010	0.0000000
1	0.1	1.0100502	0.0202010	0.0306045	0.0307606	0.0416324	0.0000000
2	0.2	1.0408108	0.0416324	0.0530813	0.0533676	0.0656507	0.0000000
3	0.3	1.0941743	0.0656504	0.0788900	0.0793533	0.0938822	0.0000000
4	0.4	1.1735108	0.0938809	0.1098406	0.1105588	0.1284070	0.0000001
5	0.5	1.2840252	0.1284025	0.1483049	0.1493995	0.1720110	0.0000002
...	...	...	...	...	...	...	...
10	1.0	2.7182699					0.0000199

It is interesting to note that the Kutta-Simpson method gives a smaller error than the modified Adams-Bashforth method for this problem. Usually this is not the case, although in some instances the Kutta-Simpson method can be expected to give superior results.

A flow chart for the numerical solution of the following example using the Kutta-Simpson method is given in Fig. 6.9.

$$y' = -\frac{x}{y^2}$$

$$y(x_0) = y_0$$

The solution is required on the range x_0 to $x_0 + mh$.

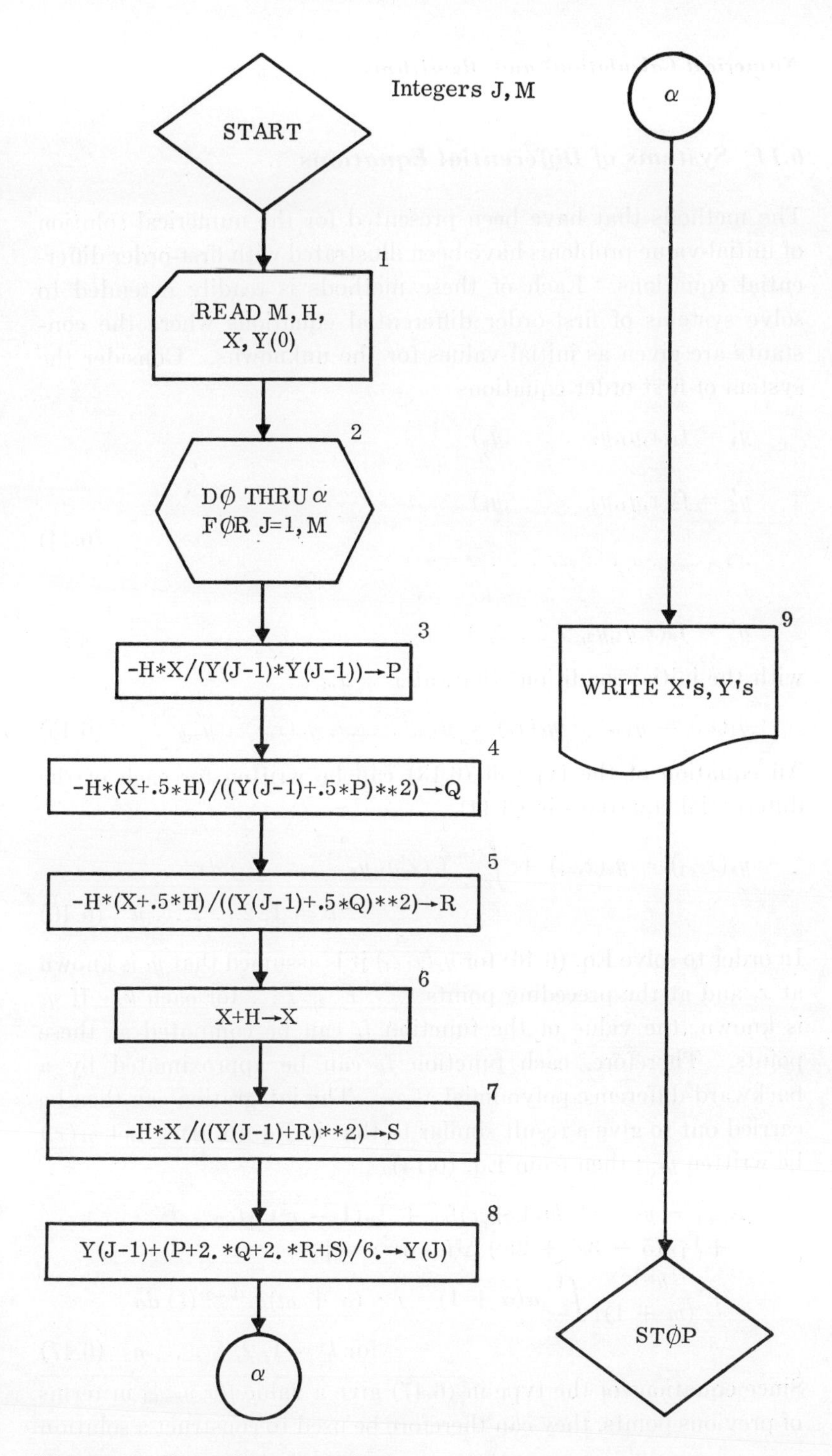

Figure 6.9 Kutta-Simpson method.

6.14 Systems of Differential Equations

The methods that have been presented for the numerical solution of initial-value problems have been illustrated with first-order differential equations. Each of these methods is readily extended to solve systems of first-order differential equations where the constants are given as initial values for the unknowns. Consider the system of first-order equations

$$\begin{aligned} y_1' &= f_1(x,y_1,y_2, \ . \ . \ . \ ,y_n) \\ y_2' &= f_2(x,y_1,y_2, \ . \ . \ . \ ,y_n) \\ &\cdots\cdots\cdots\cdots \\ y_n' &= f_n(x,y_1,y_2, \ . \ . \ . \ ,y_n) \end{aligned} \tag{6.44}$$

with the initial conditions that, at $x = x_0$,

$$y_1(x_0) = y_{1,0} \qquad y_2(x_0) = y_{2,0}, \ . \ . \ . \ , y_n(x_0) = y_{n,0} \tag{6.45}$$

An equation of the type in (6.13) can be written for each of the differential equations in (6.44).

$$y_k(x_{i+1}) = y_k(x_{i-r}) + \int_{x_{i-r}}^{x_{i+1}} f_k(x,y_1,y_2, \ . \ . \ . \ ,y_n)\, dx \qquad k = 1, 2, \ . \ . \ . \ , n \tag{6.46}$$

In order to solve Eq. (6.46) for $y_k(x_{i+1})$ it is assumed that y_k is known at x_i and at the preceding points x_{i-1}, x_{i-2}, . . . for each k. If y_k is known, the value of the function f_k can be computed at these points. Therefore, each function f_k can be approximated by a backward-difference polynomial at x_i. The integration can then be carried out to give a result similar to that in Eq. (6.14). Let $y_k(x_i)$ be written $y_{k,i}$; then from Eq. (6.14)

$$\begin{aligned} y_{k,i+1} = y_{k,i-r} &+ h[(1 + r)f_{k,i} + \tfrac{1}{2}(1 - r^2)\, \Delta f_{k,i} \\ &+ \tfrac{1}{12}(5 - 3r^2 + 2r^3)\, \Delta^2 f_{k,i} + \cdots] \\ &+ \frac{h^{m+2}}{(m + 1)!} \int_{-r}^{1} \alpha(\alpha + 1) \cdots (\alpha + m) y_k^{(m+2)}(\xi)\, d\alpha \end{aligned}$$

$$\text{for } k = 1, 2, \ . \ . \ . \ , n \tag{6.47}$$

Since equations of the type in (6.47) give a value for $y_{k,i+1}$ in terms of previous points, they can therefore be used to construct a solution to the problem, and they can also be used as a predictor formula in a predictor-corrector method.

The corrector formula for each y_k follows in a similar way from Eq. (6.21).

$$y_{k,i+1} = y_{k,i+1-r} + h\left[rf_{k,i+1} - \frac{r^2}{2}\Delta f_{k,i+1} \right.$$
$$\left. - \frac{r^2}{12}(3-2r)\,\Delta^2 f_{k,i+1} - \frac{r^2}{24}(2-r)^2\,\Delta^3 f_{k,i+1} + \cdots \right]$$
$$+ \frac{h^{m+2}}{(m+1)!}\int_{-r}^{0} \alpha(\alpha+1)\cdots(\alpha+m)y_k^{(m+2)}(\xi)\,d\alpha \qquad (6.48)$$

The Adams-Bashforth formulas for solving the system of equations in (6.44) are obtained by setting $r = 0$ in (6.47) and $r = 1$ in (6.48) and using four terms in the approximations to the integral. The resulting equations are similar to previous results except for the subscript k, which carries the solution over each equation in (6.44). The predictor and corrector formulas, respectively, for the Adams-Bashforth method are

$$p_k = y_{k,i} + \frac{h}{24}(55f_{k,i} - 59f_{k,i-1} + 37f_{k,i-2} - 9f_{k,i-3}) \qquad (6.49)$$

$$c_k = y_{k,i} + \frac{h}{24}(f_{k,i-2} - 5f_{k,i-1} + 19f_{k,i} + 9f_{k,i+1}) \qquad (6.50)$$

The procedure for solving the system of equations in (6.44) by the Adams-Bashforth formulas is illustrated by the steps to obtain the solution at $x = x_{i+1}$ when the solution at x_i, x_{i-1}, x_{i-2}, and x_{i-3} is known.

The first step in the solution is to predict the value for $y_{k,i+1}$ for each k with Eq. (6.49). This gives p_k for $k = 1, \ldots, n$. The values for p_k are then used to find an approximate value for $f_{k,i+1}$ $k = 1, \ldots, n$ at $x = x_{i+1}$. Let these values be called q_k.

$$q_k = f_k(x_{i+1},p_1,p_2, \ldots p_n) \qquad k = 1, 2, \ldots, n \qquad (6.51)$$

The respective values for q_k are then substituted for $f_{k,i+1}$ into the corrector formula (6.50) to obtain c_k. For the Adams-Bashforth predictor-corrector method, this c_k is the value for $y_{k,i+1}$.

$$y_{k,i+1} = c_k = y_{k,i} + \frac{h}{24}(f_{k,i-2} - 5f_{k,i-1} + 19f_{k,i} + 9q_k) \qquad (6.52)$$

If the modified Adams-Bashforth method is used, the final approximation for $y_{k,i+1}$ is obtained as the weighted average of c_k and p_k in

$$y_{k,i+1} = \tfrac{1}{270}(251c_k + 19p_k) \qquad k = 1, 2, \ldots, n \tag{6.53}$$

The routine is repeated to find the next point in the solution, and so on, for as many points as required.

A similar procedure can be used for any predictor-corrector method. The equations that have been developed for one first-order differential equation will carry over to a system of equations in the same way as the formulas for the Adams-Bashforth method.

The numerical solution for a system of equations using the Runge-Kutta method is also readily adapted from the previous result for one equation. The set of equations in (6.43) is modified for a system of equations as follows:

$$
\begin{aligned}
p_k &= hf_k(x_i, y_{1,i}, y_{2,i}, \ldots, y_{n,i}) \\
q_k &= hf_k\left(x_i + \frac{h}{2}, y_{1,i} + \frac{p_1}{2}, y_{2,i} + \frac{p_2}{2}, \ldots, y_{n,i} + \frac{p_n}{2}\right) \\
r_k &= hf_k\left(x_i + \frac{h}{2}, y_{1,i} + \frac{q_1}{2}, y_{2,i} + \frac{q_2}{2}, \ldots, y_{n,i} + \frac{q_n}{2}\right) \qquad (6.54) \\
s_k &= hf_k(x_i + h, y_{1,i} + r_1, y_{2,i} + r_2, \ldots, y_{n,i} + r_n) \\
y_{k,i+1} &= y_{k,} + \tfrac{1}{6}(p_k + 2q_k + 2r_k + s_k)
\end{aligned}
$$

where

$$k = 1, 2, \ldots, n$$

Equations (6.54) will give the values for each y_k at x_{i+1} when the solution at x_i is known. The first step is to find p_k for $k = 1, 2, \ldots, n$. These are then used in the computation for q_k, $k = 1, 2, \ldots, n$. Each q_k is used to compute r_k, and then all values for r_k must be known to find s_k. Each equation in (6.54) must be solved in the order listed for each k before moving on to the next step.

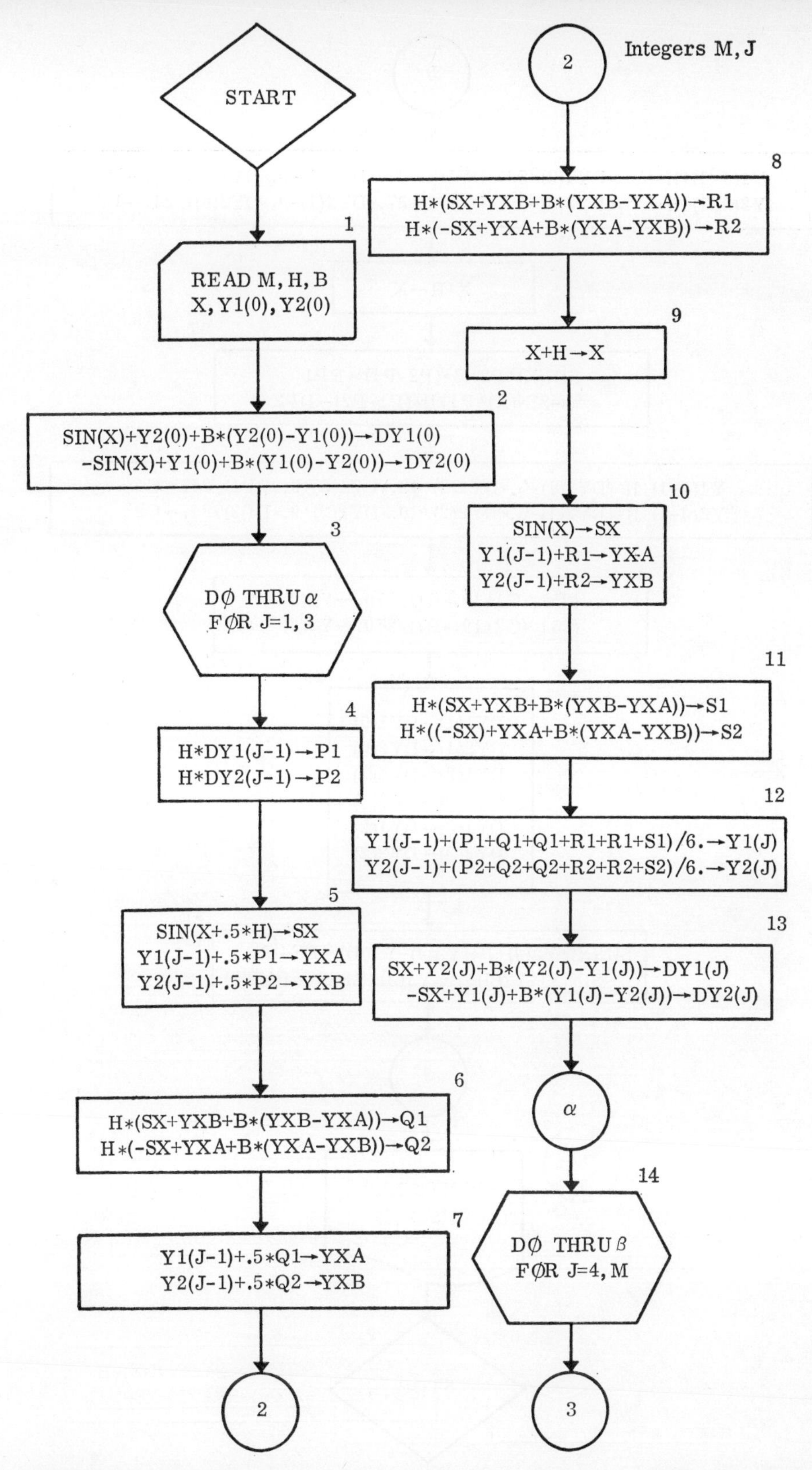

Figure 6.10 Kutta-Simpson and Adams-Bashforth methods for a system of differential equations.

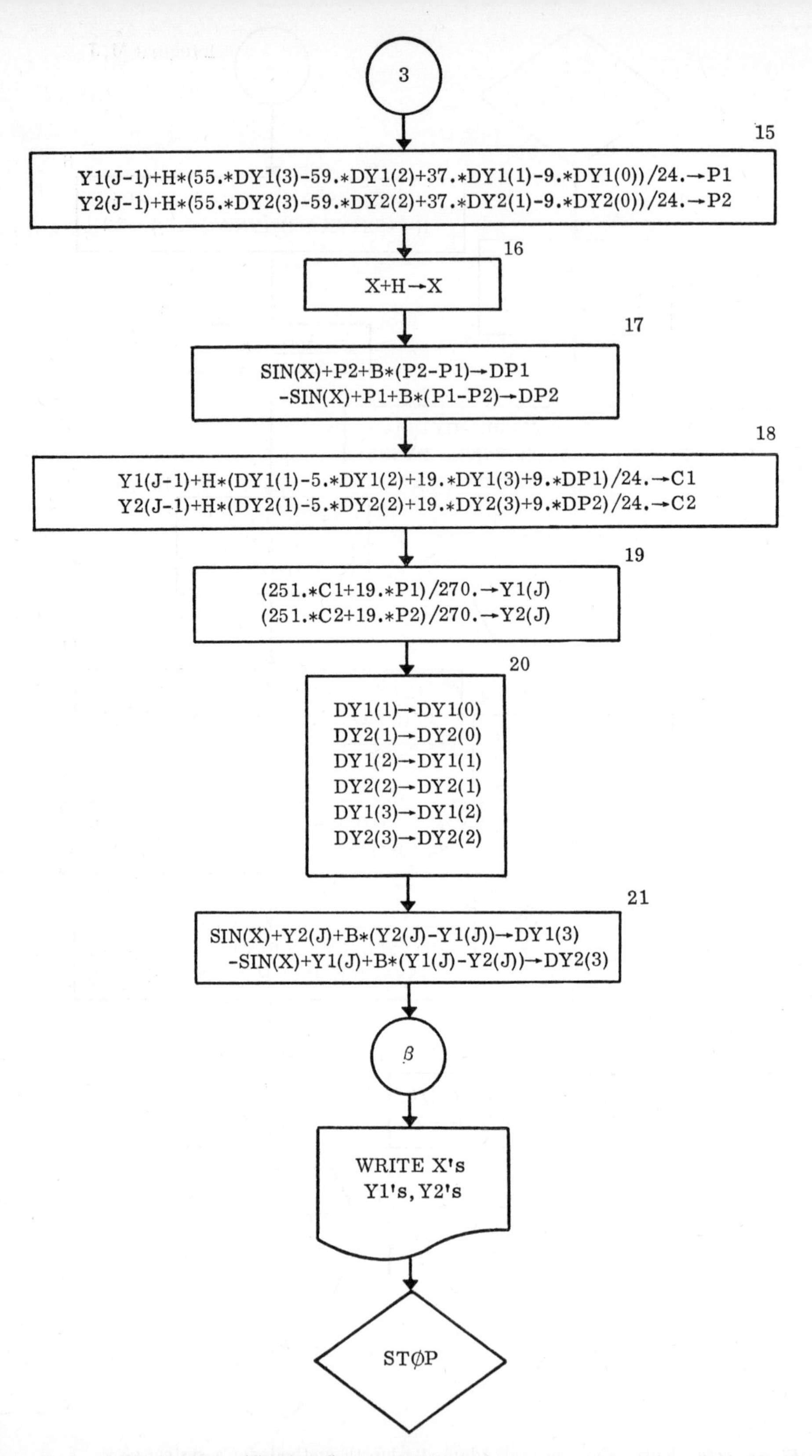

Figure 6.10 *(Continued)*

Figure 6.10 is a flow chart for solving the system of equations given in Eqs. (6.55) on the range x_0 to $x_0 + mh$.

$$y_1' = \sin x + y_2 + b(y_2 - y_1)$$

$$y_2' = -\sin x + y_1 + b(y_1 - y_2) \tag{6.55}$$

$$y_1(x_0) = y_{1,0} \qquad y_2(x_0) = y_{2,0}$$

In the flow chart the Kutta-Simpson method in Eqs. (6.54) is used to find the first three points of the solution. This is done in statements 3 to 13. The Adams-Bashforth predictor-corrector method is then used to extend the solution. Only four locations are required for each of the vectors DY1 and DY2, which represent values of the derivative. The contents of the vector must be updated in each iteration. This is done in statement 20.

6.15 Higher-order Differential Equations

The nth-order differential equation in the dependent variable y and the independent variable x is written in functional form

$$F(x,y,y',y'', \ldots ,y^{(n-1)},y^{(n)}) = 0 \tag{6.56}$$

The superscripts denote differentiation with respect to x. The necessary initial condition for this equation to have a unique solution is that at $x = x_0$ the values for y and its first $n - 1$ derivatives are assigned.

$$y(x_0) = y_0$$

$$y'(x_0) = y_0' \tag{6.57}$$

$$\cdots\cdots\cdot$$

$$y^{(n-1)}(x_0) = y_0^{(n-1)}$$

It is assumed that Eq. (6.56) can be solved explicitly for the highest derivative, $y^{(n)}$, so that it can always be written

$$y^{(n)} = f(x,y,y', \ldots ,y^{(n-1)}) \tag{6.58}$$

Equation (6.58) with the initial conditions in (6.57) can be readily transformed to a system of first-order equations. Define the variables $y_1, y_2, \ldots, y_n$ as follows:

$$
\begin{aligned}
y_1 &= y \\
y_2 &= y_1' = y' \\
y_3 &= y_2' = y'' \\
&\cdots\cdots\cdots \\
y_n &= y_{n-1}' = y^{(n-1)}
\end{aligned}
\tag{6.59}
$$

Equations (6.58) and (6.59) are arranged to give the following system of first-order equations:

$$
\begin{aligned}
y_1' &= y_2 \\
y_2' &= y_3 \\
&\cdots\cdots \\
y_{n-1}' &= y_n \\
y_n' &= f(x,y_1,y_2, \ldots ,y_n)
\end{aligned}
\tag{6.60}
$$

The system of n first-order equations in (6.60) is equivalent to the nth-order differential equation in (6.58). The initial conditions in (6.57) will transform according to

$$
\begin{aligned}
y(x_0) &= y_1(x_0) = y_{1,0} \\
y'(x_0) &= y_2(x_0) = y_{2,0} \\
y''(x_0) &= y_3(x_0) = y_{3,0} \\
&\cdots\cdots\cdots\cdots \\
y^{(n-1)}(x_0) &= y_n(x_0) = y_{n,0}
\end{aligned}
\tag{6.61}
$$

Equations (6.60) and (6.61) represent a system of first-order equations similar to Eqs. (6.44) and (6.45) and can be solved by the same numerical techniques.

Systems of higher-order differential equations can also be reduced to a system of first-order equations, as shown below.

$$\begin{aligned} \theta'' &= g(x,\theta,y,y',y'') \\ y''' &= f(x,\theta,\theta',y,y',y'') \end{aligned} \tag{6.62}$$

Let

$$\theta = y_1 \qquad \theta' = y_2$$

$$y = y_3 \qquad y' = y_4 \qquad y'' = y_5$$

Then

$$\begin{aligned} y_1' &= y_2 \\ y_2' &= g(x,y_1,y_3,y_4,y_5) \\ y_3' &= y_4 \\ y_4' &= y_5 \\ y_5' &= f(x,y_1,y_2,y_3,y_4,y_5) \end{aligned} \tag{6.63}$$

Equations (6.63) are a set of first-order equations similar to (6.44). When the appropriate initial conditions are prescribed, they can be solved by the same methods recommended for the solution of Eqs. (6.44).

6.16 Stability of the Methods for Solving the Initial-value Problem

In each of the numerical methods for solving the initial-value problem the points in the solution are in general only approximations to the true values. The errors associated with each computation will come principally from truncation of formulas and from round off of numbers in the computations. The value of the error in the computation for any one point in the solution can be controlled by the choice of the method, by the choice of the spacing h, and by the number of significant digits used in the calculations. Generally the number of digits depends upon the computer that is used, but the method and spacing are at the discretion of the programmer. Thus the error in a computation is to a considerable extent under the control of the programmer. The method and the spacing can be chosen so that the error in any computation is very small, but it generally cannot be reduced to zero.

A question of great significance is the effect of an error at one point in the solution on the subsequent results. If the effect of the error is somehow confined to the locality of the point, it will make little difference in the solution. If, on the other hand, its effect is to increase the error in each subsequent point, there is the risk that the error can overshadow the solution and the results become worthless. Should the error be propagated in such a way as to make the solution invalid, the method of solution is said to be *unstable*.

Instability can often be detected by looking at the computed results. If the computed solution appears to have a high-frequency oscillation or displays some other characteristic which the true solution should not have, the cause may be instability in the numerical computations. Should instability occur, it is essential that the computational routine be altered. Changes possible for any given problem are the choice of a different method and the choice of a different spacing h. Stability of the solution is dependent upon each of these, and the following discussion is an attempt to give some guidelines for selection of a method and a spacing.

The stability behavior of the predictor-corrector methods is dependent upon the value of the product of the spacing h and the partial derivative of the function $f(x,y)$ with respect to y. The product is designated $\bar{h}$.

$$\bar{h} = h \frac{\partial f(x,y)}{\partial y} \tag{6.64}$$

Ordinarily the user will have no control over the function $f(x,y)$ so that only h can be varied to alter $\bar{h}$. The prospect of making a change in h that will make an otherwise unstable method stable is very limited indeed. The most effective way of avoiding instability is to select a method that is stable for all expected values of $\bar{h}$. Table 6.9 gives results of stability analyses that have been made

Table 6.9
Stability Regions of Some Numerical Methods

Name of method	*Range of $\bar{h}$ for stable behavior of method*
Iterated Euler	$\bar{h} < 0$
Euler predictor-corrector	$-1.0 < \bar{h} < 0$ and $\bar{h} > 0.7$
Modified Euler predictor-corrector	$-1.5 < \bar{h} < 0$ and $\bar{h} > 0$
Adams-Bashforth	$-1.29 < \bar{h} < 0$ and $\bar{h} > 0$
Modified Adams-Bashforth	$-1.41 < \bar{h} < 0$ and $\bar{h} > 0$
Runge-Kutta	Stable for small values of $\bar{h}$

on the various methods presented in this chapter[1,2]. For each method the limiting values of $\bar{h}$ for which the method is stable are given. For example, the iterated Euler method is stable for negative values of $\bar{h}$ but is unstable for all positive values of $\bar{h}$. The modified Euler predictor-corrector method is stable for all $\bar{h} > -1.5$.

The Runge-Kutta method has been proved stable for small values of $\bar{h}$. Experience of many users indicates that it is stable even for rather large values of $\bar{h}$.

Table 6.9 should provide a guide for evaluating stability of a method. It may set limits on the size of the spacing h for a specific problem and method. For example, in Example 6.1 the value of $\bar{h} = h\partial f/\partial y = 2xh$. According to Table 6.9 the iterated Euler method will be stable only for negative values of x. The modified Adams-Bashforth formulas are stable provided $2hx > -1.41$. This places a lower limit on the product hx for stable behavior but gives a stable solution for all positive x. Clearly a method of solution may be stable for some values of the argument but unstable for others.

6.17 Comparison of Methods

On the basis of the information in Table 6.9 one would be inclined to avoid the iterated Euler and the Euler predictor-corrector methods. Each of these is unstable in a critical range for $\tilde{h}$. Of the remaining methods the one which will be most satisfactory depends upon the particular problem to be solved. The Kutta-Simpson method has the advantage of being self-starting, but it requires more computations for the derivative function. Should the derivatives be easy to evaluate, this method is a good choice. It gives a solution with a truncation error of $O(h^4)$.

The stable predictor-corrector formulas have the disadvantage of needing starting values. However, in other respects they are generally superior to the Runge-Kutta methods. Only two evaluations for the function $f(x,y)$ are needed for each point in the solution, and the error term can be explicitly defined. These are distinct advantages.

In problems where an order of approximation of $O(h^2)$ is sufficient, the modified Euler predictor-corrector method is recommended. It requires only one starting value in addition to the initial value, and this can be computed by the corrected Euler method. The programming is simple, and the method has the largest range of stability of any predictor-corrector method in Table 6.9. Should a higher-order formula be desirable, the modified Adams-Bashforth predictor-corrector method is recommended.* Its range of stability is almost as large as the modified Euler predictor-corrector method. The Adams-Bashforth method is a fourth-order method requiring four starting points, which must be approximated to $O(h^4)$. Normally this would be done with the Kutta-Simpson formulas.

6.18 Summary

This chapter is concerned with the numerical solution of initial-value problems. The initial-value problem is defined, and various methods for obtaining a numerical solution to the initial-value problem are presented. These are classified as predictor-corrector methods and single-point methods. Examples of the former are the Euler and the Adams-Bashforth predictor-corrector methods.

* See also Ref. 2 for the widely used Hamming method.

The single-point methods presented are the second- and fourth-order Runge-Kutta methods.

The expansion of these methods to obtain a numerical solution to a system of differential equations and higher-order differential equations is presented. The stability of the methods is discussed, and a summary of the results of stability studies is presented. The various methods are compared, and some recommendations for their use are made. Several flow charts outlining the various techniques are given.

Problems

1. The following set of initial-value problems is given:

(a) $y' = x^2y \quad 0 \le x \le 1$, $y(0) = 1$

(b) $y' = 2y/x \quad 1 \le x \le 2$, $y(1) = 1$

(c) $y' = y^2 + xy \quad 0 \le x \le 1$, $y(0) = -1$

(d) $y' = xy^2 \quad 0 \le x \le 1$, $y(0) = 0.1$

Use a spacing h of 0.1 and find the solution of each system by the crude Euler method.

2. Do Prob. 1 using the corrected Euler method.

3. Do Prob. 1 using the Euler predictor-corrector method. Use the corrected Euler method to find the starting value.

4. Do Prob. 1 using the modified Euler predictor-corrector method. Use the corrected Euler method to find the starting value.

5. Write a flow chart to solve the following initial-value problem by the crude Euler method for $x_i = 1 + ih$, $i = 1, 2, \ldots, m$:

$$y' = \frac{Ay}{x} \qquad 1 \le x \le 10$$

$$y(1) = 1$$

6. Do Prob. 5 using the corrected Euler method.

7. Do Prob. 5 using the Euler predictor-corrector method. Use the corrected Euler method to get the starting value.

8. Do Prob. 5 using the modified Euler predictor-corrector method. Use the corrected Euler method to get the starting value.

9. Write a flow chart to solve the following initial-value problem using the modified Euler predictor-corrector method.

$$\dot{x} = \frac{dx}{dt} = At\sqrt{x}$$

$$x(t_0) = x_0$$

Solve the problem for $t_i = t_0 + ih$, $i = 1, 2, \ldots, m$. Use the corrected Euler method for the starting value.

10. Write a flow chart to solve by the modified Euler predictor-corrector method the following initial-value problems. Use the corrected Euler method for the starting value and find the solution at $x = x_0 + ih$, $i = 1, 2, \ldots, m$.

(a) $y' = axy + z \sin x$
$z' = byz$
$y(x_0) = y_0 \qquad z(x_0) = z_0$

(b) $y' = ay - yz$
$z' = bxz$
$y(x_0) = y_0 \qquad z(x_0) = z_0$

(c) $z'' + 2bz' + z = f(x)$
$z(0) = z'(0) = 0$

(d) $v'' + \lambda v'(v^2 - 1) + v = 0$
$v(0) = 2 \qquad v'(0) = 0$

(e) $y'' = -\sin y$
$y(0) = y_0 \qquad y'(0) = 0$

11. Compare the solutions obtained by the crude Euler method at $x = 0.2$ for Prob. 1a when $h = 0.05, 0.1, 0.2$. Assume the error in the results is $O(h^k)$ and find k.

12. Do Prob. 11 with the corrected Euler method.

13. Do Prob. 1 using the Kutta-Simpson method.

14. Write a flow chart to do Prob. 1 by the Kutta-Simpson method.

15. Write a flow chart to do Prob. 9 by the Kutta-Simpson method.

16. Write a flow chart to solve the systems in Prob. 10 by the Kutta-Simpson method.

17. Solve Prob. 1a by using the Adams-Bashforth predictor-corrector method. Use the Kutta-Simpson method to find starting values.

18. Write a flow chart to solve Prob. 1 by the Adams-Bashforth method.

19. Write a flow chart to solve the systems in Prob. 10 by the modified Adams-Bashforth predictor-corrector method.

20. Write a flow chart to solve the following system using the Kutta-Simpson method:

$$Y' = xAY + B \qquad 0 \leq x \leq 1$$
$$y_i(0) = 0 \qquad i = 1, 2, \ldots, n$$

where Y and B are n-dimensional vectors and A is an n by n matrix.

References

1. Denkmann, W. John: "An Investigation of Methods for Numerically Integrating First-order Ordinary Differential Equations," M.S. thesis, University of Iowa, 1963.
2. Hamming, R. W.: "Numerical Methods for Scientists and Engineers," McGraw-Hill Book Company, New York, 1962.
3. Henrici, Peter: "Discrete Variable Method in Ordinary Differential Equations," John Wiley & Sons, Inc., New York, 1962.
4. ———: "Error Propagation for Difference Methods," John Wiley & Sons, Inc., New York, 1963.
5. ———: "Elements of Numerical Analysis," John Wiley & Sons, Inc., New York, 1964.
6. Hildebrand, F. B.: "Introduction to Numerical Analysis," McGraw-Hill Book Company, New York, 1956.
7. Levy, H., and E. A. Baggott: "Numerical Solutions of Differential Equations," Dover Publications, Inc., New York, 1934.
8. Milne, W. E.: "Numerical Solution of Differential Equations," John Wiley & Sons, Inc., New York, 1953.
9. Ralston, Anthony: "A First Course in Numerical Analysis," McGraw-Hill Book Company, New York, 1965.
10. Richtmeyer, R. D.: "Difference Methods for Initial Value Problems," Interscience Publishers, Inc., New York, 1957.
11. Scarborough, J. B.: "Numerical Mathematical Analysis," 5th ed., The Johns Hopkins Press, Baltimore, 1962.

Chapter 7
Finite Differences and Boundary-value Problems

7.1 Introduction The differential equation with the associated constraints in Eqs. (7.1) is an example of a boundary-value problem.

$$\frac{d^2y(x)}{dx^2} = y''(x) = -cx^2 \qquad 0 < x < l$$
$$y(0) = y(l) = 0 \tag{7.1}$$

The differential equation defines the behavior of the solution on the interval $0 < x < l$, and the two constraints, called *boundary conditions,* prescribe the value of the solution at each end of the interval. In general, a boundary-value problem is formulated by a differential equation on some open interval and prescribed constraints at the boundary points of the interval. This differs from the initial-value problem, where all constraints are prescribed at one value of the independent variable.

The difference in the way in which the constraints are prescribed leads to a different approach for a numerical solution to each type of problem. In the initial-value problem sufficient information is available at the starting point to construct a solution at a subsequent point. For the boundary-value problem the constraints on the solution are prescribed for different values of the variable, and sufficient information is not available at any one point to find the solution at any other point. The solution cannot be built up on a point-by-point basis, as it is in the initial-value problem. The numerical procedure for the boundary-value problem must permit the introduction of restraints at any value of the independent variable. In general, this requires that the numerical solution be constructed on the whole range of interest in the same operation. Generally this rules out the point-by-point methods of the last chapter.

There are many methods of constructing a numerical solution to a boundary-value problem. This chapter introduces the method of finite differences, which is perhaps the most elementary approach. In this method the differential equation and the boundary conditions are replaced by a set of simultaneous algebraic equations in the unknown variable by using finite-difference approximations for the derivatives. These algebraic equations can then be solved by the methods of Chap. 4 to obtain an approximation to the solution of the boundary-value problem.

7.2 *Finite-difference Approximations for Derivatives*

Finite-difference approximations for the derivatives of a function of x at $x = x_i$ are obtained by using the Taylor-series expansion of the function about $x = x_i$. This series is

$$f(x) = f(x_i + \Delta x) = f(x_i) + \Delta x\, f'(x_i) + \frac{1}{2!}\Delta x^2 f''(x_i) + \frac{1}{3!}\Delta x^3 f'''(x_i) + \cdots \tag{7.2}$$

If Δx is set equal to h and then to $-h$, the following two equations are obtained:

$$f(x_i + h) = f(x_i) + hf'(x_i) + \frac{1}{2!}h^2 f''(x_i) + \frac{1}{3!}h^3 f'''(x_i) + O(h^4)$$

$$f(x_i - h) = f(x_i) - hf'(x_i) + \frac{1}{2!}h^2 f''(x_i) - \frac{1}{3!}h^3 f'''(x_i) + O(h^4) \tag{7.3}$$

An approximation of $O(h^2)$ is found for the first derivative of the function if one of Eqs. (7.3) is subtracted from the other.

$$f'(x_i) = \frac{1}{2h}[f(x_i + h) - f(x_i - h)] + O(h^2) \tag{7.4}$$

When the two equations are added together, an $O(h^2)$ approximation is found for the second derivative.

$$f''(x_i) = \frac{1}{h^2}[f(x_i + h) - 2f(x_i) + f(x_i - h)] + O(h^2) \tag{7.5}$$

Equations (7.4) and (7.5) are called the central-difference approximations for the first and second derivatives of the function $f(x)$ at $x = x_i$. If the two equations for $f(x)$ at $x = x_i + 2h$ and $x = x_i - 2h$ are included with Eqs. (7.3), it is easily shown that the $O(h^2)$ approximations to the third and fourth derivatives of $f(x)$

at $x = x_i$ are given by

$$f'''(x_i) = \frac{1}{2h^3}[f(x_i + 2h) - 2f(x_i + h) + 2f(x_i - h) - f(x_i - 2h)] + O(h^2) \quad (7.6)$$

$$f^{\mathrm{IV}}(x_i) = \frac{1}{h^4}[f(x_i + 2h) - 4f(x_i + h) + 6f(x_i) - 4f(x_i - h) + f(x_i - 2h)] + O(h^2) \quad (7.7)$$

Equations (7.6) and (7.7) are the central-difference approximations to the third and fourth derivatives of $f(x)$ at $x = x_i$. Central-difference approximations to higher-order derivatives can be obtained by the same general approach.

By using the central-difference approximations for the derivative terms in a differential equation it is possible to reduce the differential equation to a set of algebraic equations in discrete values of the variable.

7.3 *Finite-difference Analog*

The finite-difference analog for a boundary-value problem is a set of algebraic equations in discrete values of the unknown functions. The set of algebraic equations is obtained by (1) replacing the differential equation at prescribed points by an approximate-difference equation and (2) specifying the boundary conditions in terms of the discrete values of the unknown variables introduced in (1). For purpose of explanation the finite-difference analog is developed for several different examples of boundary-value problems, and a numerical solution is obtained for some of the examples.*

Example 7.1

The boundary-value problem in Eqs. (7.1) is solved by finite differences.

$$\begin{aligned} y'' &= -cx^2 \qquad 0 < x < l \\ y(0) &= y(l) = 0 \end{aligned} \quad (7.1)$$

* See references listed at the end of this chapter for more complete discussion.

The interval $0 \leq x \leq l$ is divided into n equal parts of length h, as shown in Fig. 7.1.

$$x_i = ih \qquad i = 0, 1, 2, \ldots, n \tag{7.8}$$

The end points of the spaces at $x = x_i$ are called *nodal points;* the points on the open interval $0 < x < l$ are called *interior points;* and the points at $x = x_0$ and $x = x_n$ are called *boundary points.* At each interior point of the interval the central-difference approximation for

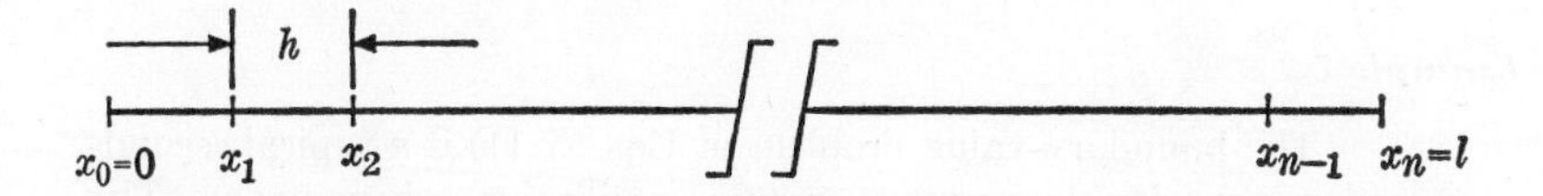

Figure 7.1 Nodal spacing for $x_i = ih$.

the differential equation is written. From the definition of the second derivative in Eq. (7.5) the finite-difference analog for the differential equation in (7.1) at $x = x_i$ is

$$y_{i-1} - 2y_i + y_{i+1} = -h^2cx_i^2 \qquad i = 1, 2, \ldots, n-1 \tag{7.9}$$

Equation (7.9) represents $n - 1$ equations in the $n + 1$ discrete values for y at $x = x_i$, $i = 0, 1, \ldots, n$. The two boundary conditions give two more equations in the unknowns, so that a system of $n + 1$ equations in $n + 1$ unknowns is defined.

$$y_0 = 0$$

$$y_{i-1} - 2y_i + y_{i+1} = -h^2cx_i^2 \qquad i = 1, 2, \ldots, n-1 \tag{7.10}$$

$$y_n = 0$$

Equations (7.10) make up the finite-difference analog for the problem in Eqs. (7.1). The truncation error in the difference formula for the derivative is $O(h^2)$, and the solution of Eqs. (7.10) is an $O(h^2)$ approximation to the true solution. A reduction of h by one-half would reduce the truncation error to approximately one-fourth its previous value.

Table 7.1 gives numerical results from Eqs. (7.10) when c and l have values of 12 and 1, respectively.

The approximate solution when $h = \frac{1}{2}$ and $h = \frac{1}{4}$ is tabulated under $n = 2$ and $n = 4$, respectively, and the exact solution is given in the last column of the table. Note that the error in the solution at the midpoint, when $h = \frac{1}{2}$, is four times the error when $h = \frac{1}{4}$.

Table 7.1

x	y when $n = 2$	y when $n = 4$	Exact y
0.00	0.000000	0.000000	0.000000
0.25		0.234375	0.246094
0.50	0.375000	0.421875	0.437500
0.75		0.421875	0.433594
1.00	0.000000	0.000000	0.000000

Example 7.2

The boundary-value problem in Eqs. (7.11) is a typical second-order equation governing certain oscillation phenomena. The functions $f(x)$ and $g(x)$ are assumed known.

$$y'' + f(x)y = g(x) \qquad 0 < x < l$$
$$y(0) = y_0 \qquad y(l) = y_n \tag{7.11}$$

The difference analog for Eqs. (7.11) is obtained by writing the central-difference approximation for the differential equation at prescribed nodal points on the open interval, $0 < x < l$, and using the given boundary conditions. Let the nodal points be prescribed in the same way as Example 7.1. Then at $x = x_i$

$$y_{i-1} - 2y_i + y_{i+1} + h^2 f(x_i) y_i = h^2 g(x_i) \qquad i = 1, 2, \ldots, n-1 \tag{7.12}$$

If this set of $n - 1$ equations is combined with the boundary conditions, the following set of $n + 1$ equations in $n + 1$ unknowns is obtained.

$$\begin{aligned}
&y_0 = y_0\\
&y_0 + (h^2 f_1 - 2)y_1 + y_2 = h^2 g_1\\
&y_1 + (h^2 f_2 - 2)y_2 + y_3 = h^2 g_2\\
&\cdots\cdots\cdots\cdots\cdots\cdots\\
&y_{n-2} + (h^2 f_{n-1} - 2)y_{n-1} + y_n = h^2 g_{n-1}\\
&y_n = y_n
\end{aligned} \tag{7.13}$$

The notation $f_i = f(x_i)$ and $g_i = g(x_i)$ is used in (7.13) and throughout this chapter. A solution to this set of equations for the y_i's is the $O(h^2)$ approximation to the exact solution of the problem.

In order to solve the system of algebraic equations obtained in the examples, the methods of Chap. 4 are employed. Let the system of equations in Eqs. (7.13) be represented by

$$\sum_{j=0}^{n} a_{ij}y_j = a_{i,n+1} \qquad i = 0, 1, 2, \ldots, n \tag{7.14}$$

Then

$$a_{00} = 1 \qquad a_{0,n+1} = y_0$$

$$a_{i,i-1} = a_{i,i+1} = 1 \qquad a_{ii} = h^2 f_i - 2 \qquad a_{i,n+1} = h^2 g_i \qquad i = 1, 2, \ldots, n-1 \tag{7.14a}$$

$$a_{nn} = 1 \qquad a_{n,n+1} = y_n$$

and all other coefficients in the matrix are zero.

The computational procedure for finding the elements of the matrix in Eqs. (7.14*a*) is given in the flow chart in Fig. 7.2.

Table 7.2
Numerical Solutions to Example 7.2

x	$f(x)$	$g(x)$	y_5 ($n = 5$)	y_{10} ($n = 10$)	y_{20} ($n = 20$)
0.0000000	0.0000000	1.0000000	0.0000000	0.0000000	0.0000000
0.0500000	0.0500000	0.9500000			−0.0158743
0.1000000	0.1000000	0.9000000		−0.0293760	−0.0293739
0.1500000	0.1500000	0.8500000			−0.0406150
0.2000000	0.2000000	0.8000000	−0.0497500	−0.0497230	−0.0497158
0.2500000	0.2500000	0.7500000			−0.0567919
0.3000000	0.3000000	0.7000000		−0.0619700	−0.0619574
0.3500000	0.3500000	0.6500000			−0.0653264
0.4000000	0.4000000	0.6000000	−0.0671000	−0.0670310	−0.0670133
0.4500000	0.4500000	0.5500000			−0.0671332
0.5000000	0.5000000	0.5000000		−0.0658240	−0.0658025
0.5500000	0.5500000	0.4500000			−0.0631396
0.6000000	0.6000000	0.4000000	−0.0593800	−0.0592890	−0.0592649
0.6500000	0.6500000	0.3500000			−0.0543013
0.7000000	0.7000000	0.3000000		−0.0483970	−0.0483744
0.7500000	0.7500000	0.2500000			−0.0416129
0.8000000	0.8000000	0.2000000	−0.0342400	−0.0341660	−0.0341484
0.8500000	0.8500000	0.1500000			−0.0261156
0.9000000	0.9000000	0.1000000		−0.0176630	−0.0176522
0.9500000	0.9500000	0.5000000			−0.0088992
1.0000000	1.0000000	0.0000000	0.0000000	0.0000000	0.0000000

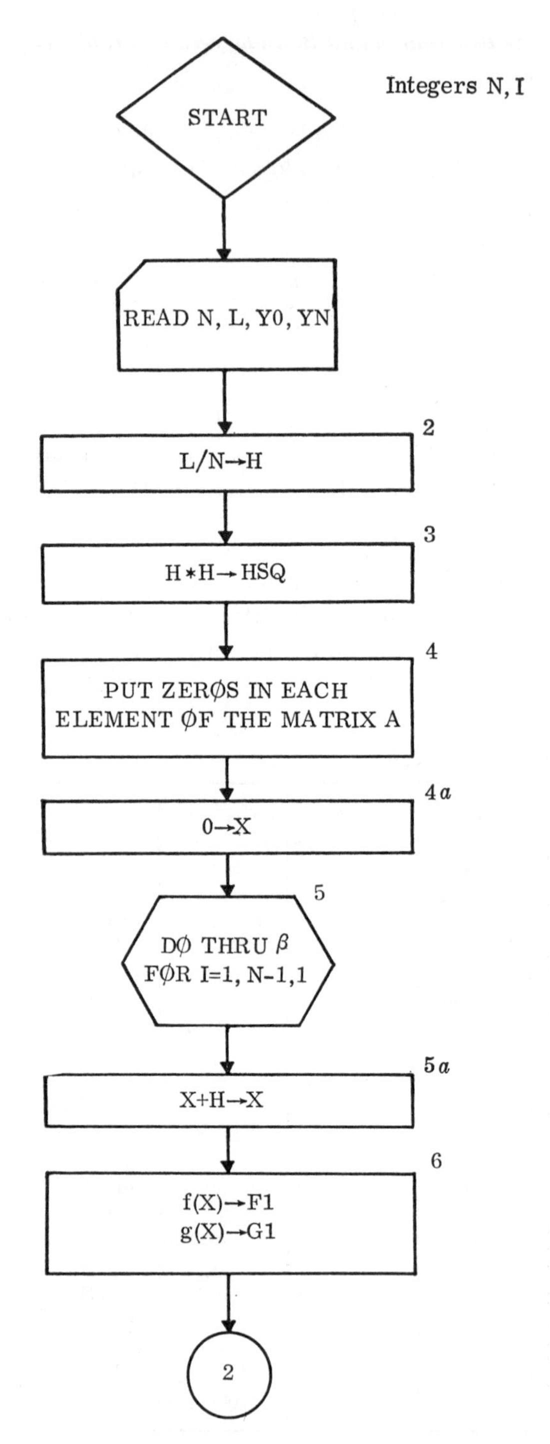

Figure 7.2 Generation of matrix for a finite-difference analog.

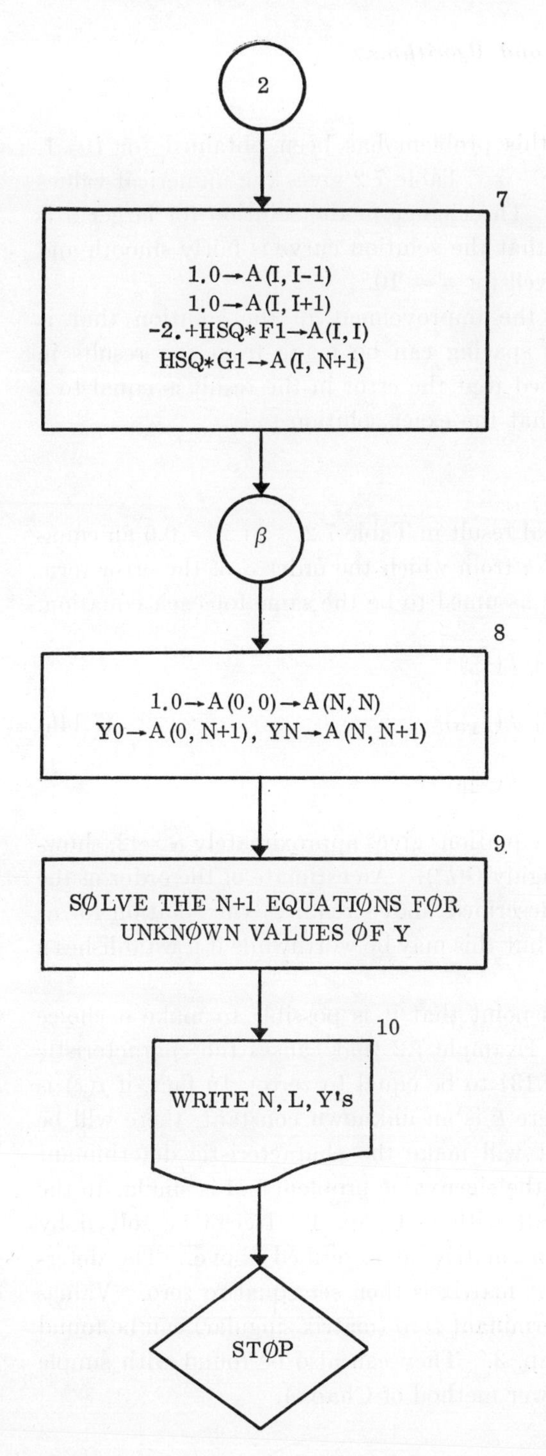

Figure 7.2 *(Continued)*

The solution for this problem has been obtained for $l = 1$, $f(x) = x$, and $g(x) = 1 - x$. Table 7.2 gives the numerical values for $n = 5$, 10, and 20. The change in the solution for larger n is small, which indicates that the solution curve is fairly smooth and is approximated very well for $n = 10$.

An evaluation of the improvement in the solution that is obtained for a smaller spacing can be made from the results in Table 7.2. It is assumed that the error in the result is equal to a constant times h^α, so that the exact solution y_e is

$$y_e = y_a + kh^\alpha$$

where y_a is the numerical result in Table 7.2. At $x = 0.6$ an equation is written for each h from which the order α of the error term can be estimated. k is assumed to be the same for each equation.

$$\begin{aligned} y_e &= -0.0593800 + k(1/5)^\alpha \\ y_e &= -0.0592890 + k(1/10)^\alpha \\ y_e &= -0.0592649 + k(1/20)^\alpha \end{aligned} \tag{7.14b}$$

A solution of this set of equations gives approximately $\alpha = 2$, showing that the error is roughly $O(h^2)$. An estimate of the order of the error along the lines described above requires the solution for at least three values of h, but this may be worthwhile if it establishes a satisfactory result.

It is noted at this point that it is possible to make a choice of the function $f(x)$ in Example 7.2 that causes the characteristic determinant of Eqs. (7.13) to be equal to zero. In fact, if $f(x)$ is replaced by $kf_1(x)$, where k is an unknown constant, there will be $n - 1$ values of k that will make the characteristic determinant equal to zero. This is the eigenvalue problem and is similar to the eigenvalue problem dealt with in Chap. 4. It can be solved by setting up the coefficient matrix, as described above. The determinant of the coefficient matrix is then set equal to zero. Values of k that make the determinant zero (matrix singular) can be found by the methods of Chap. 3. They can also be found with simple modification by the power method of Chap. 4.

7.4 Difference Analog for Higher-order Differential Equations

The finite-difference analog for an ordinary differential equation up to and including the fourth order is easily obtained by using Eqs. (7.4), (7.5), (7.6), and (7.7).

The boundary conditions that are prescribed for differential equations of order greater than 2 will almost always involve derivatives of the solution at the boundaries of the interval. When the boundary conditions are given in terms of derivatives of the solution, it is necessary to approximate the boundary conditions with finite differences. These difference approximations are a part of the difference analog for the problem. An example will illustrate the central-difference analog for a fourth-order equation when some of the boundary conditions are given in terms of derivatives.

Example 7.3

$$y^{\text{IV}} + f(x)y''' + g(x)y = p(x) \qquad 0 < x < l$$

$$y(0) = y_0 \qquad y(l) = y_n \tag{7.15}$$

$$y'(0) = y_0 \qquad y''(l) = y_n''$$

The interval $0 \le x \le l$ is divided into n spaces of length h, so that

$$x_i = ih \qquad i = 0, \ldots, n \tag{7.8}$$

The derivatives in Eqs. (7.15) are replaced by the $O(h^2)$ difference approximations to give

$$\left(1 - \frac{h}{2}f_i\right) y_{i-2} - (4 - hf_i)y_{i-1} + (6 + h^4 g_i)y_i - (4 + hf_i)y_{i+1} + \left(1 + \frac{h}{2}f_i\right) y_{i+2} = h^4 p_i \tag{7.16}$$

This equation will hold at each of the interior points on the interval, i.e., at $i = 1, 2, \ldots, n - 1$ (see Fig. 7.3).

When $i = 1$, Eq. (7.16) includes a value for y at $x = -h$, and when $i = n - 1$, it includes a value for y at $x = l + h$. These are unknowns outside the region of interest but are brought into the problem by the difference analog for the differential equation. It is necessary to extrapolate the solution curve a distance h beyond each end of the interval in order to include these two unknowns. Figure 7.4 shows each end of the interval $0 \le x \le l$ and how the solution curve might be extrapolated beyond the end points. The properties of the solution curve at the end points are determined by the boundary conditions. The conditions at $x = 0$ prescribe the value for y_0

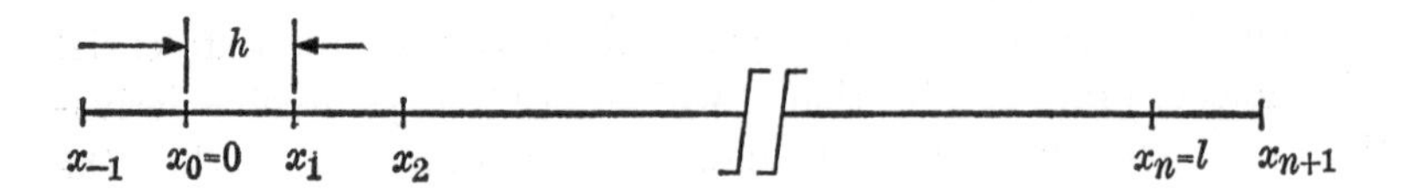

Figure 7.3 Nodal spacing for higher-order equation when $x_i = ih$.

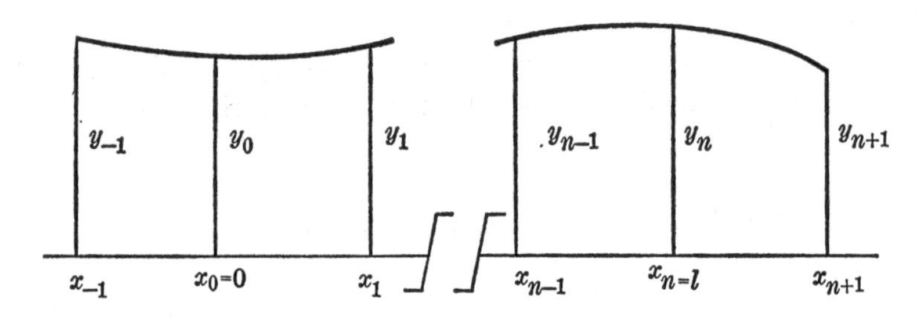

Figure 7.4 Boundary conditions.

and the slope of the curve through y_0. The central-difference $O(h^2)$ approximation for the slope at y_0 is given in terms of y_1 and y_{-1}.

$$y_0' = \frac{1}{2h}(y_1 - y_{-1}) + O(h^2) \tag{7.17}$$

At $x = l$ the value for $y(l) = y_n$ and the second derivative are given. The central-difference $O(h^2)$ approximation for the second derivative at $x = l$ is

$$y_n'' = \frac{1}{h^2}(y_{n+1} - 2y_n + y_{n-1}) + O(h^2) \tag{7.18}$$

Equations (7.17) and (7.18) and the given values for y_0 and y_n are four additional independent equations in the unknown values of y. These four equations, with the $n - 1$ equations in (7.16), give a set of $n + 3$ algebraic equations that can be solved for the $n + 3$ unknowns. This set of $n + 3$ equations is the central-difference analog for the boundary-value problem in Eqs. (7.15). They are written out below for clarity and can be solved by any of the methods given in Chap. 4.

$$-y_{-1} + y_1 = 2hy_0'$$

$$y_0 = y_0$$

$$\left(1 - \frac{h}{2}f_i\right) y_{i-2} - (4 - hf_i)y_{i-1} + (6 + h^4g_i)y_i - (4 + hf_i)y_{i+1} + \left(1 + \frac{h}{2}f_i\right) y_{i+2} = h^4p_i \quad \text{for } i = 1, 2, \ldots, n-1 \tag{7.19}$$

$$y_n = y_n$$

$$y_{n-1} - 2y_n + y_{n+1} = h^2y_n''$$

The finite-difference analog in Eqs. (7.19) has a truncation error of $O(h^2)$. Improvement in the approximation is made by decreasing the spacing h. However, there are distinct limitations to this, which will be discussed later in the chapter.

The examples have illustrated how a central-difference analog is constructed for an ordinary differential equation on a given interval. In the examples the interval is first divided into equal spaces by nodal points. At each nodal point that is on the open interval (an interior point) the differential equation is replaced by the $O(h^2)$ central-difference approximation. The analog that is obtained will always involve unknown values of the variable at nodal points which are outside the open interval. Thus, the difference analog for the differential equation will contain more unknowns than equations. The additional equations that are needed for a unique solution come from the boundary conditions in the problem. In example 7.1 the difference analog includes the boundary values y_0 and y_n which are prescribed by the two end conditions. The difference analog for Example 7.3 brings four unknowns that are outside the open interval $0 < x < l$ into the equations. The four boundary conditions give the four additional equations necessary for a solution. In general the difference approximations for the boundary conditions must contain only those unknowns which are included in the analog for the differential equation. This condition is necessary if the system of algebraic equations is to have a unique solution. In addition, if the approximations are to have truncation error of the same order throughout, the boundary conditions must be written to the same order of approximation as the differential equation. These two restrictions will sometimes impose some modification of the spacing used in the previous examples. For example, consider the following boundary-value problem.

Example 7.4

$$y'' + f(x)y = g(x) \qquad 0 < x < l$$
$$y'(0) = y_0' \qquad y(l) = y_n \tag{7.20}$$

This problem differs from Example 7.2 only in the boundary conditions. Here the first derivative of y is prescribed at $x = 0$ rather than the displacement. Let the spacing on the interval be equal to the spacing given in Fig. 7.1. The finite-difference analog for the differential equation is then

$$y_{i-1} + (-2 + h^2 f_i)y_i + y_{i+1} = h^2 g_i \qquad \text{for } i = 1, 2, \ldots, n-1 \tag{7.21}$$

This gives a set of $n - 1$ equations in the $n + 1$ unknowns y_i, $i = 0, 1, \ldots, n$. The central-difference approximation for the derivative at $x = 0$ will involve an extrapolated point on the solution curve.

$$y_0' = \frac{1}{2h}(y_1 - y_{-1}) + O(h^2) \tag{7.22}$$

The extrapolated point y_{-1} is not included in the difference analog for the differential equation. Therefore, Eq. (7.22) is not admissible as a part of the finite-difference analog.

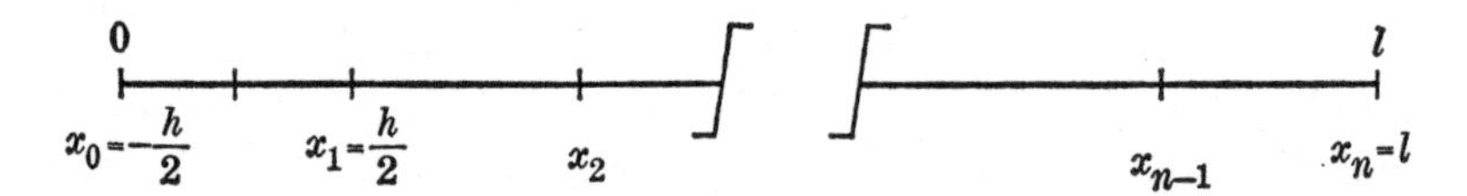

Figure 7.5 Nodal spacing for $x_i = (l - \frac{1}{2})h$.

A central-difference approximation for the boundary condition that will use only the unknowns in the analog for the differential equation can be obtained by altering the spacing geometry. Figure 7.5 shows a spacing geometry where the left end of the interval marked 0 falls at the midpoint of the space x_0 to x_1. The nodal points are defined in

$$x_i = (i - \tfrac{1}{2})h \qquad \text{for } i = 0, 1, 2, \ldots, n \tag{7.23}$$

The relationship between the spacing h, the number of spaces n, and the length of the interval l is

$$l = (n - \tfrac{1}{2})h \tag{7.24}$$

The set of equations in (7.21) will hold at each interior point of the interval, $i = 1, 2, \ldots, n - 1$, and will include the extrapolated point at $x = x_0 = -h/2$. The central-difference approximation of $O(h^2)$ for the derivative at $x = 0$ is (see Eqs. 7.32 for $l = 0$)

$$y'(0) = \frac{1}{h}(y_1 - y_0) + O(h^2) \tag{7.25}$$

This equation involves only the unknowns in Eq. (7.21). When combined with them and the condition that $y = 0$ at $x = l$, a set of $n + 1$ equations in $n + 1$ unknowns is obtained.

$$\begin{aligned}
-y_0 + y_1 &= hy'(0) \\
y_0 - (2 - h^2f_1)y_1 + y_2 &= h^2g_1 \\
y_1 - (2 - h^2f_2)y_2 + y_3 &= h^2g_2 \\
\cdots\cdots\cdots\cdots&\cdots\cdots \\
y_{n-2} - (2 - h^2f_{n-1})y_{n-1} + y_n &= h^2g_{n-1} \\
y_n &= y(l)
\end{aligned} \tag{7.26}$$

The displacement at the left boundary is not obtained directly from this analog but is readily approximated. An $O(h^2)$ approximation for the deflection at $x = 0$ is given by taking an average of the values to each side.

$$y(0) = \tfrac{1}{2}(y_0 + y_1) + O(h^2) \tag{7.27}$$

Example 7.5

The differential equation and the boundary conditions in Eqs. (7.28) arise in structural analysis.

$$\begin{aligned}
&y^{\mathrm{IV}} + f(x)y = g(x) \qquad 0 < x < l \\
&y(0) = y'(0) = 0 \qquad y''(l) = y'''(l) = 0
\end{aligned} \tag{7.28}$$

The spacing geometry is chosen so that the boundary at $x = 0$ is at a nodal point and the boundary at $x = l$ is at the midpoint of a node (see Fig. 7.6). The nodal points are given by

$$x_i = ih \qquad i = -1, 0, 1, \ldots, n, n+1$$
$$l = x_n - \frac{h}{2} = h(n - \tfrac{1}{2}) \tag{7.29}$$

The differential equation is replaced by the central-difference analog

$$y_{i-2} - 4y_{i-1} + (6 + h^4 f_i)y_i - 4y_{i+1} + y_{i+2} = h^4 g_i \qquad \text{for } i = 1, 2, \ldots, n-1 \tag{7.30}$$

which holds at each of the nodal points on the interval $0 < x < l$

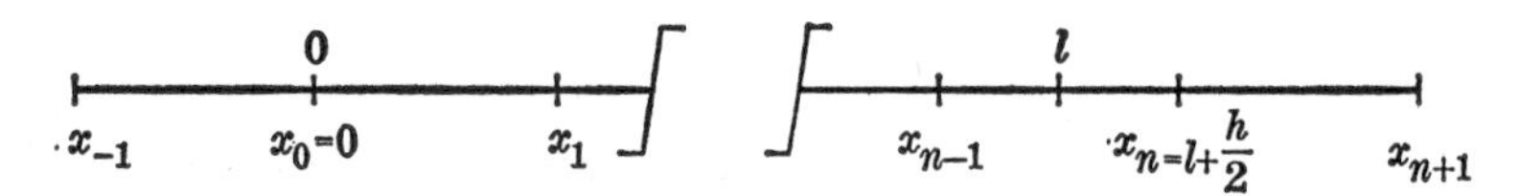

Figure 7.6 Nodal spacing for $x_i = ih$ and $x_n = l + h/2$.

and gives a set of $n - 1$ equations in the $n + 3$ unknowns $y_{-1}, y_0, y_1, \ldots, y_n, y_{n+1}$. The boundary conditions at $x = 0$ are

$$-y_{-1} + y_1 = 2hy'(0) = 0$$
$$y_0 = y(0) = 0 \tag{7.31}$$

At $x = l$ it is necessary to define the second and third derivatives at the midpoint of a space in terms of the unknowns at the nodal spaces.

If the Taylor-series expansion in Eq. (7.2) is made about $x = l$, which is the midpoint of a nodal space, the function at $x = l + h/2$ and $x = l - h/2$ is defined by

$$f(l + \tfrac{1}{2}h) = f(l) + \tfrac{1}{2}hf'(l) + \frac{1}{2!}\frac{h^2}{4}f''(l) + \frac{1}{3!}\frac{h^3}{8}f'''(l) + O(h^4)$$
$$f(l - \tfrac{1}{2}h) = f(l) - \tfrac{1}{2}hf'(l) + \frac{1}{2!}\frac{h^2}{4}f''(l) - \frac{1}{3!}\frac{h^3}{8}f'''(l) + O(h^4) \tag{7.32}$$

Clearly the $O(h^2)$ approximation to the first derivative at $x = l$ is given by subtracting the second of Eqs. (7.32) from the first. If the series for the function is written for $x = l + \tfrac{3}{2}h$ and $x = l - \tfrac{3}{2}h$, it can easily be verified that the $O(h^2)$ approximations for the second and third derivatives at $x = l$ are

$$f''(l) = \frac{1}{2h^2}[f(l + \tfrac{3}{2}h) - f(l + \tfrac{1}{2}h) - f(l - \tfrac{1}{2}h) + f(l - \tfrac{3}{2}h)] + O(h^2) \quad (7.33)$$

$$f'''(l) = \frac{1}{h^3}[f(l + \tfrac{3}{2}h) - 3f(l + \tfrac{1}{2}h) + 3f(l - \tfrac{1}{2}h) - f(l - \tfrac{3}{2}h)] + O(h^2) \quad (7.34)$$

The approximations in Eqs. (7.33) and (7.34) give the central-difference formulas for the boundary conditions at $x = l$. These are combined with Eqs. (7.30) and (7.31) to give the difference analog for Example 7.5.

$$-y_{-1} + y_1 = 0$$

$$y_0 = 0$$

$$y_{i-2} - 4y_{i-1} + (6 + h^4 f_i)y_i - 4y_{i+1} + y_{i+2} = h^4 g_i \qquad \text{for } i = 1, 2, \ldots, n-1 \quad (7.35)$$

$$y_{n-2} - y_{n-1} - y_n + y_{n+1} = 0$$

$$-y_{n-2} + 3y_{n-1} - 3y_n + y_{n+1} = 0$$

Equations (7.35) are a set of $n + 3$ equations in $n + 3$ unknowns and can be solved by one of the methods of Chap. 4.

It is noted that for certain choices of the function f the determinant of the coefficient matrix in Eqs. (7.35) will go to zero. Such a possibility should not be ignored in the numerical solution.

7.5 Solution of the Difference Analog

The finite-difference analog for a boundary-value problem is a set of simultaneous algebraic equations in discrete values of the unknown variable. If the differential equation and boundary conditions are linear in the unknown, the algebraic equations are also linear, and the methods of Chap. 4 can be used to obtain the solution of the set of equations.

In general, two methods are used to solve the simultaneous equations. One is the Gauss-Seidel iteration, or some modification of it, and the other is the Gaussian elimination. In each method the number of computations is greatly reduced by utilizing the fact that all the nonzero entries in the coefficient matrix are grouped in a narrow band about the principal diagonal. In Example 7.2 they occupy the three spaces on and to each side of the main diagonal. Example 7.3 has but five nonzero coefficients in each equation that are on and adjacent to the main diagonal of the coefficient matrix. The difference analog for any boundary-value problem exhibits these same characteristics, and this property can be used to great advantage for the solution of the unknowns.

An elimination method of solution for the finite-difference analog in Example 7.2 which utilizes the high density of zeros in the coefficient matrix is explained in some detail. The jth equation of the difference analog is

$$r_{j1}y_{j-1} + r_{j2}y_j + r_{j3}y_{j+1} = b_j \tag{7.36}$$

The coefficients r_{jk} are the nonzero coefficients in the set of equations (7.13) and can be evaluated by comparing Eqs. (7.36) and (7.13). The first four equations of the analog are written out for clarity.

$$\begin{aligned}
r_{12}y_1 + r_{13}y_2 \qquad\qquad\qquad\qquad &= b_1 \\
r_{21}y_1 + r_{22}y_2 + r_{23}y_3 \qquad\qquad &= b_2 \\
r_{31}y_2 + r_{32}y_3 + r_{33}y_4 \qquad &= b_3 \\
r_{41}y_3 + r_{42}y_4 + r_{43}y_5 &= b_4
\end{aligned} \tag{7.37}$$

The coefficient r_{11} is the coefficient on y_0 and is not included in Eqs. (7.37) because y_0 is a given value. The constant term $r_{11}y_0$ is moved to the right-hand side of the equation as a part of b_1.

The first step in the solution of the analog is to eliminate y_1 from the set of equations in (7.37). The first equation is divided by r_{12} to give

$$y_1 + r_{13}^{(1)}y_2 = b_1^{(1)} \tag{7.38}$$

where

$$r_{13}^{(1)} = \frac{r_{13}}{r_{12}} \quad \text{and} \quad b_1^{(1)} = \frac{b_1}{r_{12}}$$

Only two divisions are necessary in this operation, compared with the n division operations in the general method, because the remaining coefficients in the first equation are all equal to zero.

The unknown y_1 is eliminated from the second of Eqs. (7.37) to give

$$r_{22}^{(1)}y_2 + r_{23}y_3 = b_2^{(1)} \tag{7.39}$$

where

$$r_{22}^{(1)} = r_{22} - r_{21}r_{13}^{(1)} \quad \text{and} \quad b_2^{(1)} = b_2 - r_{21}b_1^{(1)}$$

Again, the number of computations has been reduced because of the zeros in the equations. The unknown y_1 does not appear in any of the remaining equations, and no further elimination operations need be performed.

The steps in Eqs. (7.38) and (7.39) can now be repeated on the second and third equations to eliminate y_2. This sequence of operations is repeated until y_{n-2} is eliminated from the $(n - 1)$st equation (the value for y_n is given as a boundary condition and is treated similarly to y_0). The operations that are performed to eliminate y_j from the $(j + 1)$st equation are

$$\begin{aligned} r_{j3}^{(1)} &= \frac{r_{j3}}{r_{j2}} \qquad b_j^{(1)} = \frac{b_j}{r_{j2}} \\ r_{j+1,2}^{(1)} &= r_{j+1,2} - r_{j+1,1}r_{j,3}^{(1)} \qquad b_{j+1}^{(1)} = b_{j+1} - r_{j+1,1}b_j^{(1)} \end{aligned} \tag{7.40}$$

After the elimination has been performed for $j = n - 2$, the $(n - 1)$st equation is of the form

$$r_{n-1,2}^{(1)}y_{n-1} = b_{n-1}^{(1)} \tag{7.41}$$

Solving for y_{n-1} gives

$$y_{n-1} = \frac{b_{n-1}^{(1)}}{r_{n-1,2}^{(1)}} \tag{7.42}$$

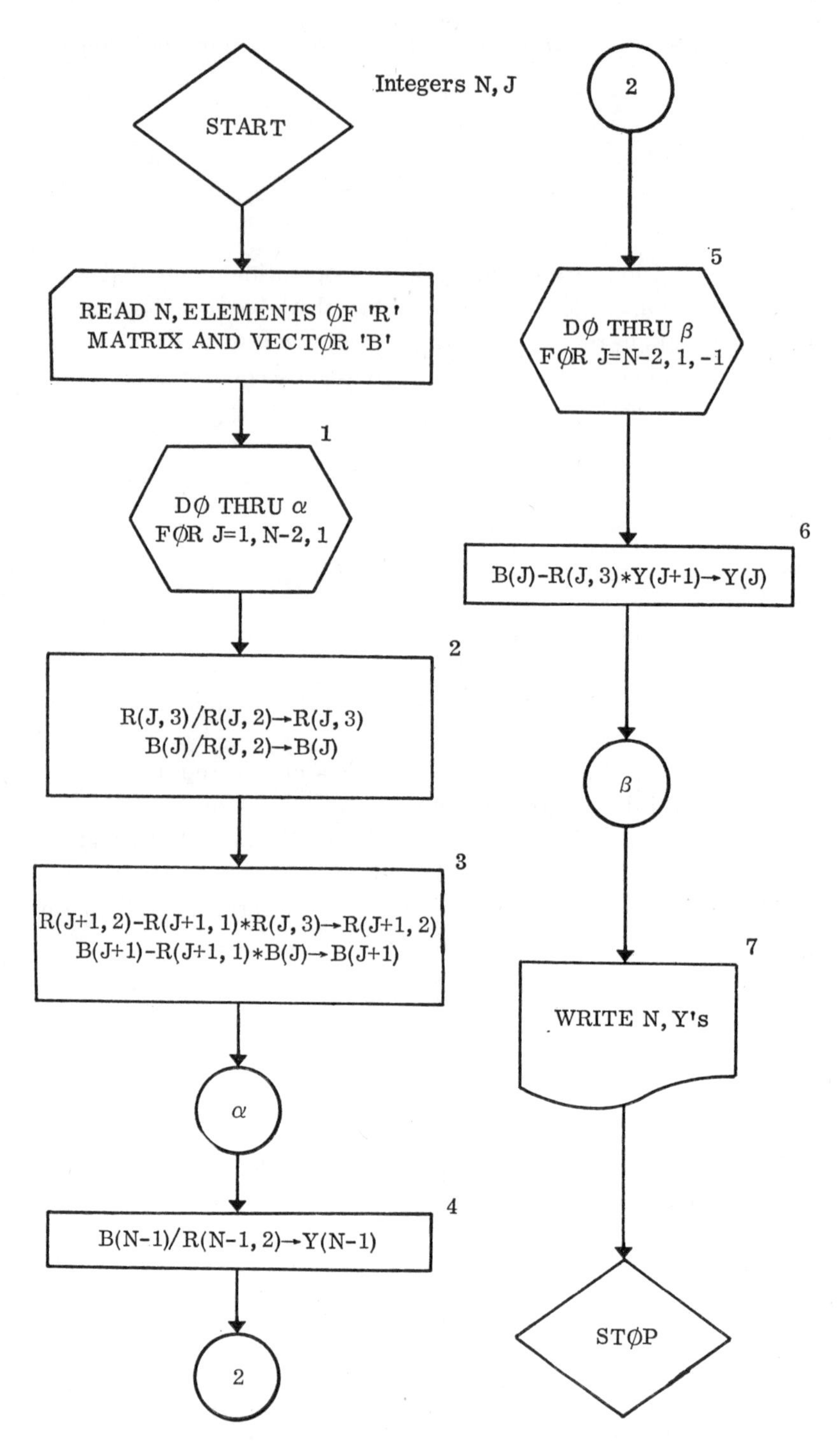

Figure 7.7 Gaussian elimination to solve difference analog for second-order equations.

The jth equation in the set is of the form

$$y_j + r_{j,3}^{(1)} y_{j+1} = b_j^{(1)} \tag{7.43}$$

which is solved for y_j.

$$y_j = b_j^{(1)} - r_{j,3}^{(1)} y_{j+1} \tag{7.44}$$

Equation (7.44) is used to solve for the unknowns y_j in reverse order, i.e., for $j = n - 2, n - 3, \ldots, 2, 1$.

Figure 7.7 is a flow chart to solve the set of $n - 1$ equations in (7.36) according to the procedure outlined above.

A general procedure is now developed for solving the difference analog containing the unknowns y_j, $j = 1, 2, \ldots, n$, by both the Gauss-Seidel iteration and by Gaussian elimination. The jth equation of a difference analog can be written

$$r_{j,1}y_{j-k} + r_{j,2}y_{j-k+1} + \cdots + r_{j,k+1}y_j + \cdots + r_{j,2k+1}y_{j+k} = b_j \tag{7.45}$$

The coefficients on y are the nonzero coefficients in the finite-difference analog. The value of k is fixed by the order of the differential equation. For Example 7.1, $k = 1$, and for Example 7.3, $k = 2$.

In the Gauss-Seidel iteration method for solving the system of equations represented by Eq. (7.45), which is discussed first, each equation of the set is solved for y_j.

$$y_j = (b_j - r_{j,1}y_{j-k} - r_{j,2}y_{j-k+1} \cdots - r_{j,k}y_{j-1} - r_{j,k+2}y_{j+1} \cdots - r_{j,2k+1}y_{j+k})/r_{j,k+1} \tag{7.46}$$

for $j = 1, 2, \ldots, n$. Starting values are assumed for each unknown, and the iteration is performed until there is no significant change in the value of any unknown.

Figure 7.8 is a flow chart for the Gauss-Seidel iteration using Eq. (7.46). In this flow chart values for the nonzero coefficients are assumed known and are read into the memory of the machine. The number of equations n, the bandwidth parameter k, and an error term used for the iteration routine are also read into memory. The flow chart is written to solve the unknowns $y_1, y_2, \ldots, y_n$. Minor modifications can be made if there are more or less than n unknowns in the analog. Statements 1 and 2 generate starting values for the iteration routine. Statement 4 is the control for

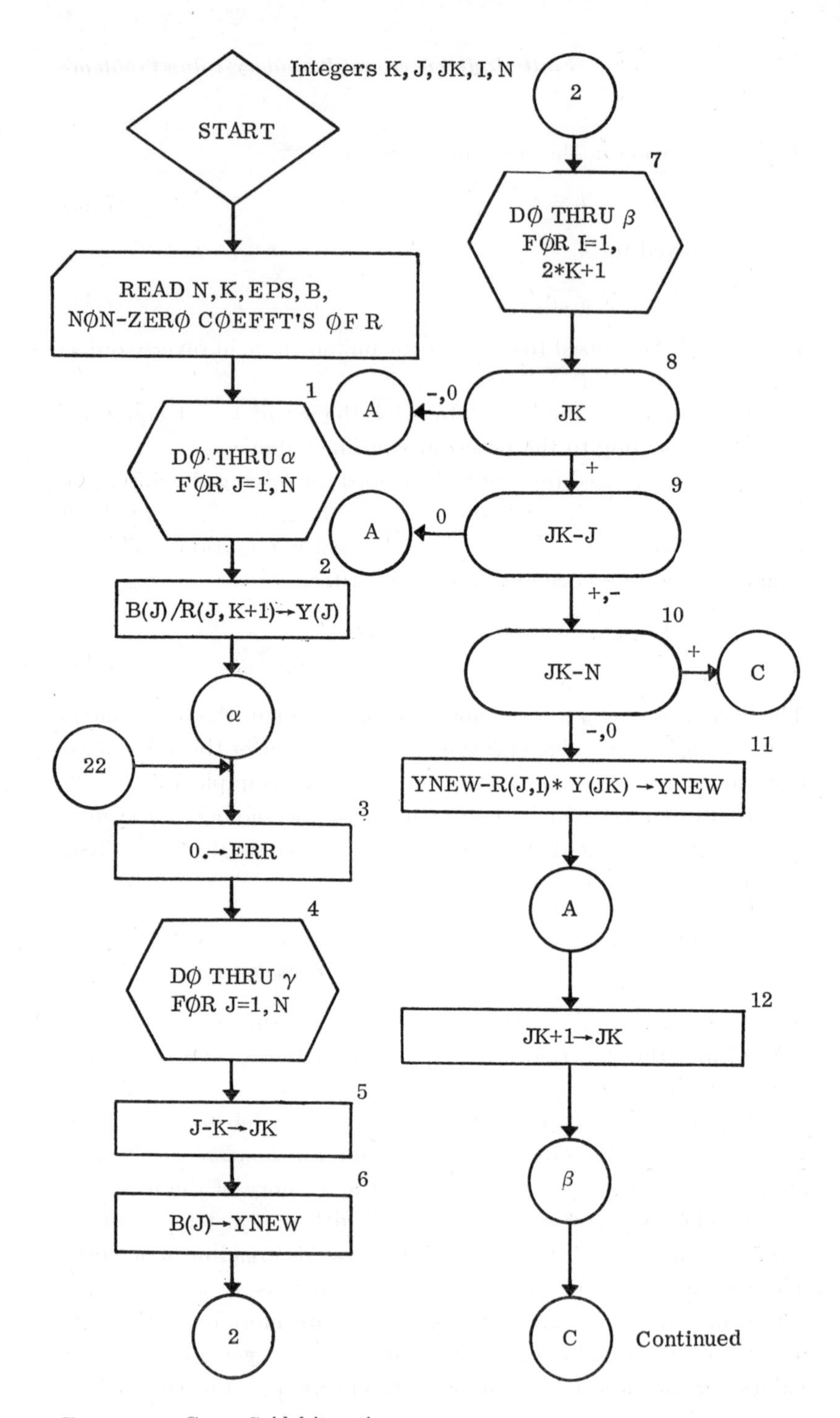

Figure 7.8 Gauss-Seidel iteration.

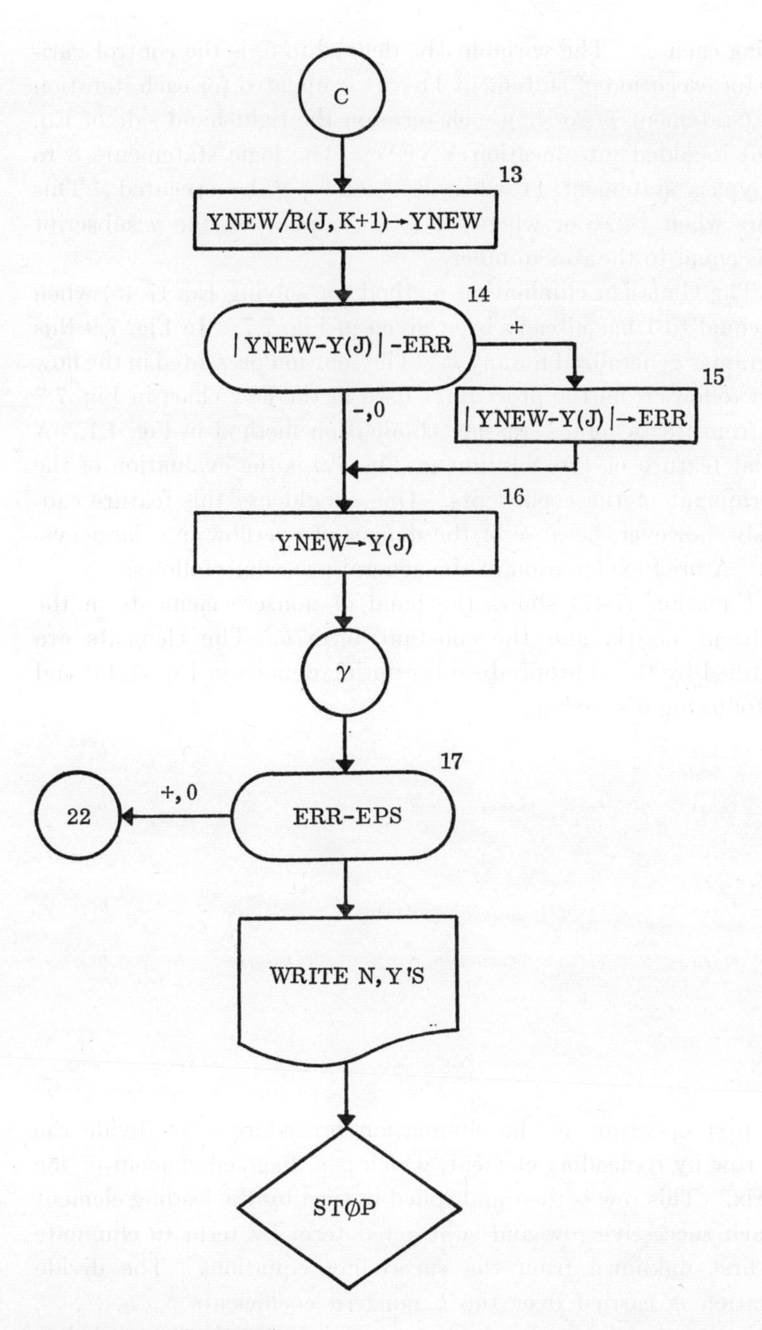

Figure 7.8 *(Continued)*

finding each y_j. The variable JK, defined in 5, is the control variable for execution of statement 11. It is updated for each iteration on i (statement 7) so that each term on the right-hand side of Eq. (7.46) is added into location YNEW. The logic statements 8 to 10 bypass statement 11 when it should not be executed. This occurs when $j < k$ or when $j + k > n$ and when the y subscript JK is equal to the row number.

The Gaussian elimination method for solving Eq. (7.45) when k is equal to 1 has already been given in Fig. 7.7. In Fig. 7.9 this program is generalized for any k. The solution presented in the flow chart follows from the procedures used in the flow chart in Fig. 7.7 and from the general Gaussian elimination method in Fig. 4.1. A special feature of the solution in Fig. 7.9 is the evaluation of the determinant of the coefficients. One should use this feature cautiously, however, because of the danger of overflow in a large system. A brief explanation of the general procedure follows.

Equation (7.47) shows the band of nonzero elements in the coefficient matrix and the constant term b. The elements are identified by the appropriate subscripting defined in Eq. (7.45) and the following discussion.

$$
\begin{array}{cccccccccccc}
r_{1,k+1} & r_{1,k+2} & \cdots & r_{1,2k+1} & & & & & & & & b_1 \\
r_{2,k} & r_{2,k+1} & \cdots & r_{2,2k} & r_{2,2k+1} & & & & & & & b_2 \\
\cdots & \cdots & \cdots & \cdots & \cdots & \cdots & \cdots & \cdots & \cdots & \cdots & \cdots & \cdots \\
r_{k+1,1} & r_{k+1,2} & \cdots & r_{k+1,k+1} & & \cdots & r_{k+1,2k+1} & \cdots & & & & b_{k+1} \\
& r_{k+2,1} & \cdots & r_{k+2,k} & r_{k+2,k+1} & & \cdots & & r_{k+2,2k+1} & \cdots & & b_{k+2} \\
\cdots & \cdots & \cdots & \cdots & \cdots & \cdots & \cdots & \cdots & \cdots & \cdots & \cdots & \cdots \\
& & & & & r_{n,1} & r_{n,2} & \cdots & & r_{n,k+1} & & b_n
\end{array}
\tag{7.47}
$$

The first operation in the elimination procedure is to divide the first row by its leading element, which is a diagonal element of the matrix. This row is then multiplied in turn by the leading element of each successive row and subtracted term by term to eliminate the first unknown from the succeeding equations. The divide operation is carried over the k nonzero coefficients $r_{1,k+2}, \ldots,$ $r_{1,2k+1}$, and the elimination is performed only for the k rows below the first. For the elimination on the second column the second row is divided by the current value of its diagonal element, which is $r_{2,k+1}$. Elimination of the k elements in the second column below the diagonal is made in a way similar to the elimination on the elements of the first column.

Consider now the operations for placing zeros in the column position below the diagonal element of the jth row (elimination of the jth unknown). Previous operations have eliminated the first $j - 1$ unknowns, so that there is a zero in each of the first $j - 1$ column positions for row $j, j + 1, \ldots, n$. Divide row j by the diagonal element, $r_{j,k+1}$. This involves k divisions for the coefficient matrix when $n - j \geq k$ and only $n - j$ divisions when $n - j < k$. This can be verified by reference to Eq. (7.47). For $j \leq n - k$ there are k nonzero elements to the right of the diagonal; whereas, for $j > n - k$ there are but $n - j$ terms to the right of the diagonal in the coefficient matrix. The elements in the jth column below the diagonal are eliminated by using the jth row. This involves the elimination of k terms when $n - j \geq k$ and only $n - j$ terms when $n - j < k$. Again this is verified by reference to Eq. (7.47).

The operations described above are carried out for $j = 1, 2, \ldots, n - 1$, which gives a coefficient matrix with unity in each diagonal location except the last and zero in each position below the diagonal. The solution for the unknowns can then be carried out by backward substitution. Again only the nonzero elements in the matrix are used.

The flow chart (Fig. 7.9) for performing the elimination described above and solving for the unknowns by back substitution is explained below.

1,2. The first element in the coefficient matrix is placed in DET, where the value of the determinant will be stored, and control is set for successive elimination of the first $n - 1$ unknowns.

4. The limit on the iteration statements in the elimination operations that follow is set. If $j + k$ is less than or equal to n, there are k nonzero coefficients to the right of the diagonal term in the jth equation. These coefficients are $r_{j,k+2}, r_{j,k+3}, \ldots, r_{j,2k+1}$, and the elimination must be carried over these k terms. Also there are k equations immediately below equation j in which the jth unknown must be eliminated. Thus when $j + k \leq n$, then k iterations on each subscript must be made. If $j + k > n$, then the limits

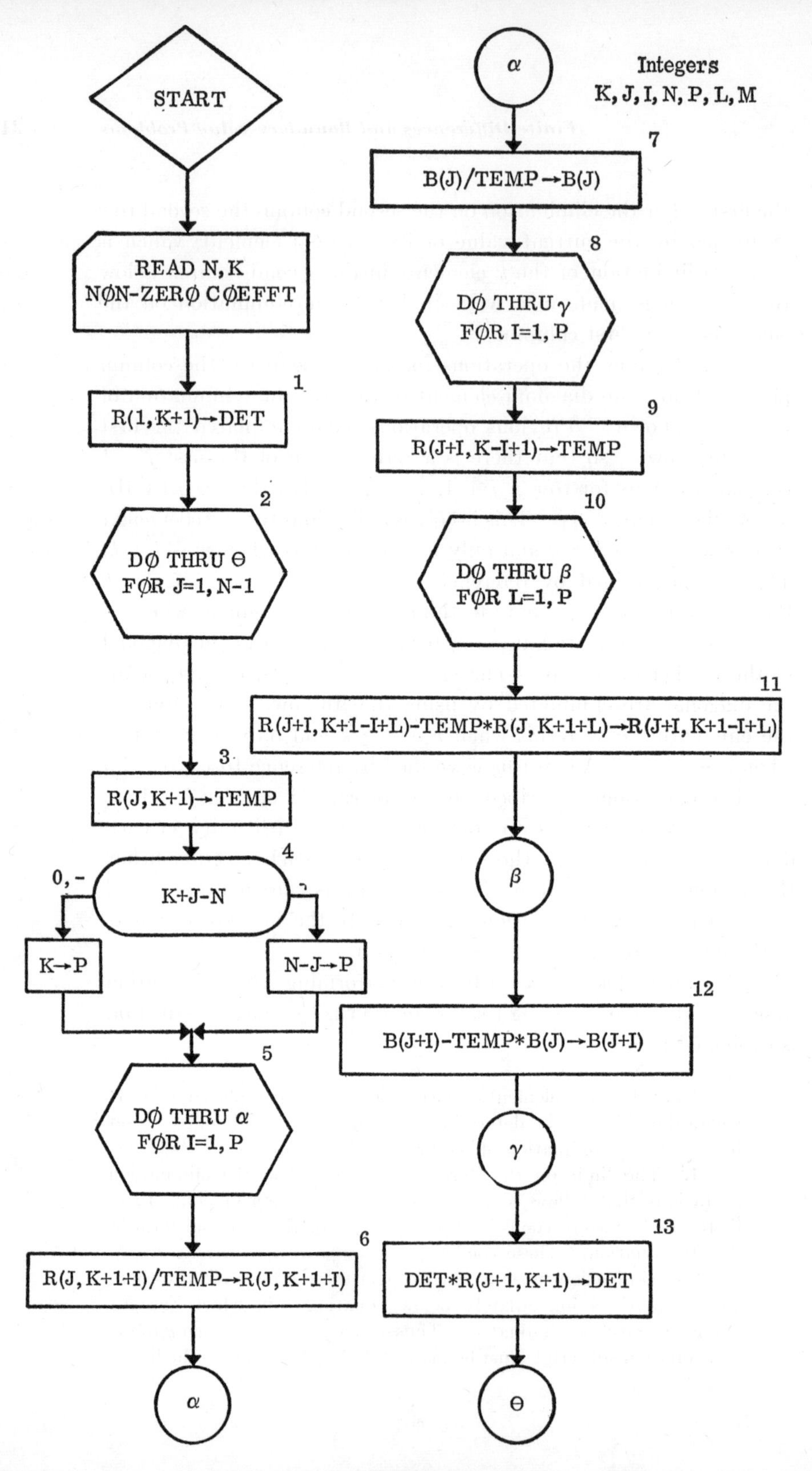

Figure 7.9 Modified Gaussian elimination.

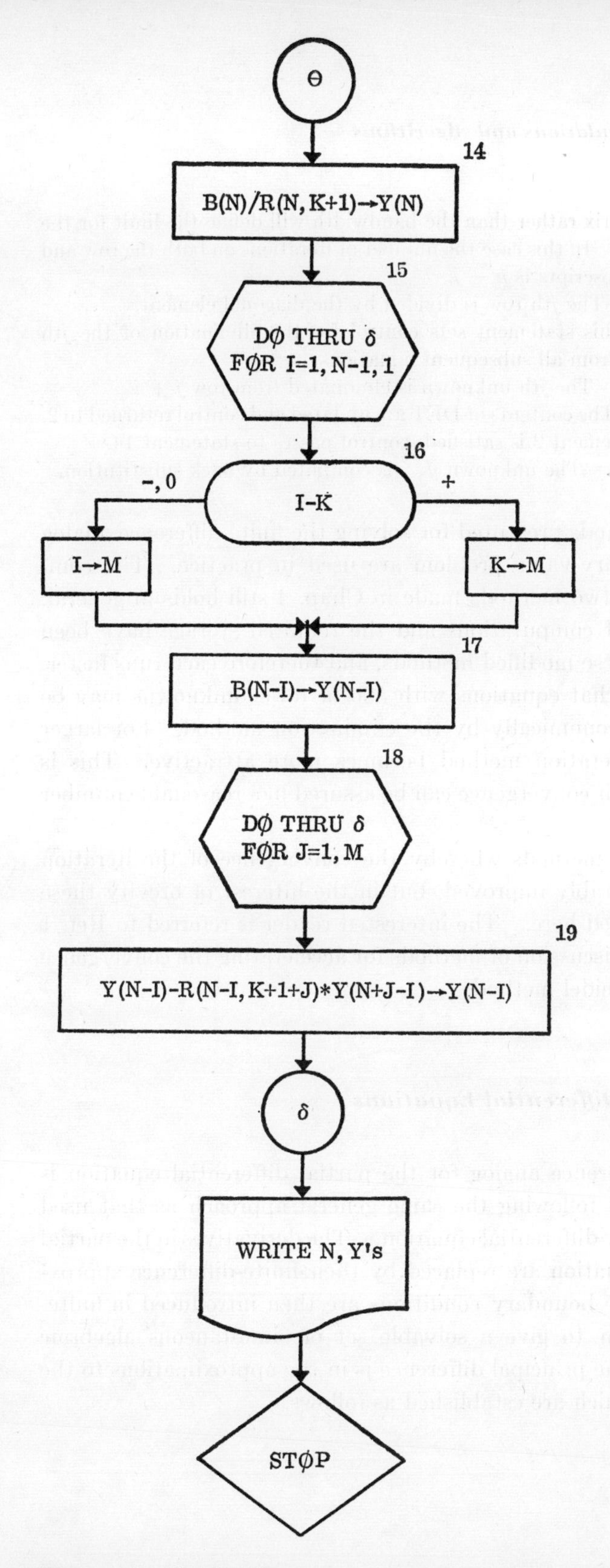

Figure 7.9 (*Continued*)

of the matrix rather than the bandwidth will define the limit for the iterations. In this case the number of iterations on both the row and column subscripts is $n - j$.

5–7. The jth row is divided by the diagonal element.

8. This statement sets control for the elimination of the jth unknown from all subsequent equations.

9–12. The jth unknown is eliminated from row $j + i$.

13. The contents of DET are updated and control returned to 2. When statement 2 is satisfied, control passes to statement 14.

14–19. The unknown y_{n-i} is computed by back substitution.

Both methods presented for solving the finite-difference analog for the boundary-value problem are used in practice. The comparison of the two methods made in Chap. 4 still holds in general. The number of computations and the required storage have been cut down in these modified methods, and therefore each runs faster. It is still true that equations with 500 or fewer unknowns may be solved more economically by the elimination method. For larger systems the iteration method becomes more attractive. This is especially true if convergence can be assured in a reasonable number of iterations.

There are methods whereby the convergence of the iteration can be considerably improved, but in the interest of brevity these are not discussed here. The interested reader is referred to Ref. 1 for a detailed discussion of methods for accelerating the convergence of the Gauss-Seidel method.

7.6 Partial Differential Equations

The finite-difference analog for the partial differential equation is constructed by following the same general approach as that used for the ordinary differential equation. The derivatives in the partial differential equation are replaced by their finite-difference approximations. The boundary conditions are then introduced in finite-difference form to give a solvable set of simultaneous algebraic equations. The principal difference is in the approximations to the derivatives, which are established as follows.

Let the function $f(x,y)$ be a function of the two independent variables x and y. In order to find the difference approximations to the partial derivatives the Taylor series for a function of two variables is used. The procedure is similar to that used for finding approximations to ordinary derivatives.

Equation (7.48) is the Taylor series for $f(x,y)$ expanded about the point (x_i,y_j). The partial derivatives in the formula are evaluated at (x_i,y_j).

$$\begin{aligned} f(x,y) = f(x_i + \Delta x, y_j + \Delta y) &= f(x_i,y_j) + \Delta x \frac{\partial f}{\partial x} + \Delta y \frac{\partial f}{\partial y} \\ &+ \frac{1}{2!} \Delta x^2 \frac{\partial^2 f}{\partial x^2} + \Delta x \, \Delta y \frac{\partial^2 f}{\partial x \, \partial y} + \frac{1}{2!} \Delta y^2 \frac{\partial^2 f}{\partial y^2} + \frac{1}{3!} \Delta x^3 \frac{\partial^3 f}{\partial x^3} \\ &+ \frac{1}{2!} \Delta x^2 \, \Delta y \frac{\partial^3 f}{\partial x^2 \, \partial y} + \frac{1}{2!} \Delta x \, \Delta y^2 \frac{\partial^3 f}{\partial x \, \partial y^2} \\ &+ \frac{1}{3!} \Delta y^3 \frac{\partial^3 f}{\partial y^3} + \cdots \end{aligned} \qquad (7.48)$$

The difference approximations for the partial derivatives with respect to x are obtained from Eq. (7.48) by first setting Δy equal to zero. The $O(h^2)$ approximations for the first and second partial derivatives with respect to x are then found from the two equations that are obtained when Δx is set equal to plus and minus h.

$$\frac{\partial f(x_i,y_j)}{\partial x} = \frac{1}{2h} [f(x_i + h, y_j) - f(x_i - h, y_j)] + O(h^2) \qquad (7.49)$$

$$\frac{\partial^2 f(x_i,y_j)}{\partial x^2} = \frac{1}{h^2} [f(x_i + h, y_j) - 2f(x_i,y_j) + f(x_i - h, y_j)] + O(h^2) \qquad (7.50)$$

In a similar way the partial derivatives with respect to y are found.

$$\frac{\partial f(x_i,y_j)}{\partial y} = \frac{1}{2k} [f(x_i, y_j + k) - f(x_i, y_j - k)] + O(k^2) \qquad (7.51)$$

$$\frac{\partial^2 f(x_i,y_j)}{\partial y^2} = \frac{1}{k^2} [f(x_i, y_j + k) - 2f(x_i,y_j) + f(x_i, y_j - k)] + O(k^2) \qquad (7.52)$$

Figure 7.10 shows the nodal points that are used to find the derivative approximations. The point designated i, j has coordinates (x_i, y_j). The spacing between the nodal lines in the x direction is h and in the y direction k. The partial derivatives at x_i, y_j with respect to x are all defined for j fixed, and the partial derivatives with respect to y are defined for i fixed. The cross-derivative term is obtained from the Taylor-series expansion by setting Δx and Δy in Eq. (7.48) equal to the combinations of values (h,k), $(h,-k)$,

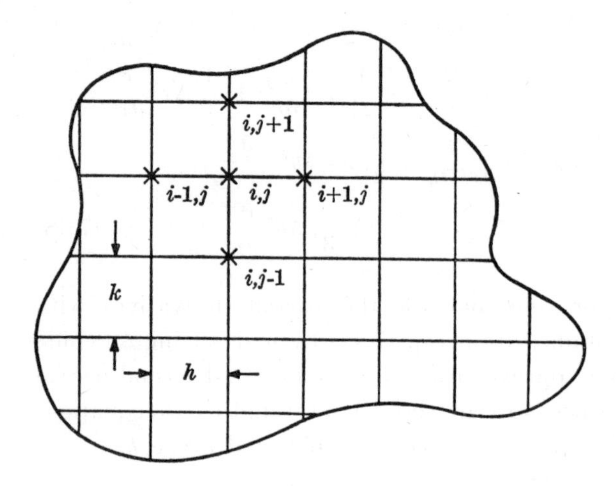

Figure 7.10 Nodal points for partial derivative approximations.

$(-h,k)$, $(-h,-k)$ and solving the resulting four equations for $\partial^2 f/(\partial x\, \partial y)$.

$$\frac{\partial^2 f(x_i,y_j)}{\partial x\, \partial y} = \frac{1}{4hk}\left[f(x_i + h,\, y_j + k) - f(x_i - h,\, y_j + k) - f(x_i + h,\, y_j - k) + f(x_i - h,\, y_j - k)\right] + O(h - k) + O(hk) \quad (7.53)$$

Note that the truncation error is $O(hk)$ only when $h = k$. Higher-order derivatives are easily constructed by using the Taylor series but are not presented here. The reader is referred to standard texts on numerical methods for a detailed discussion of this problem and a tabulation of difference approximations for higher derivatives.

The application of the finite-difference method to a partial differential equation is illustrated in the following example.

Example 7.6

Find the finite-difference analog for

$$\frac{\partial^2 u}{\partial x^2} + \frac{\partial^2 u}{\partial y^2} = 0 \qquad 0 < x < a,\ 0 < y < b \tag{7.54}$$

$$u(0,y) = f_1(y) \qquad u(a,y) = f_2(y)$$

$$u(x,0) = g_1(x) \qquad u(x,b) = g_2(x)$$

The functions f_1, f_2, g_1, and g_2 are known functions of their argument. They represent the value of the function u on the corresponding boundary. The differential equation in (7.54) is Laplace's equation and is often written as $\nabla^2 u = 0$, where

$$\nabla^2 = \frac{\partial^2}{\partial x^2} + \frac{\partial^2}{\partial y^2}$$

is the Laplacian operator.

In order to obtain the finite-difference analog for the problem, the rectangular space $0 \le x \le a$, $0 \le y \le b$ is divided into the network shown in Fig. 7.11. The lines $x =$ const are spaced a distance h apart and are labeled x_i, $i = 0, 1, \ldots, m$. The lines $y =$ const are spaced a distance k apart and are labeled y_j, $j = 0, 1, \ldots, n$. The finite-difference analog for the differential equation at the point (x_i,y_j) is obtained by replacing the derivatives in the equation with the finite-difference approximations given in Eqs. (7.50) and (7.52).

$$\frac{1}{h^2}(u_{i+1,j} - 2u_{i,j} + u_{i-1,j}) + \frac{1}{k^2}(u_{i,j+1} - 2u_{i,j} + u_{i,j-1}) = 0 \tag{7.55}$$

Equation (7.55) is an algebraic equation in the unknown values of u and holds at each interior point in the region $i = 1, \ldots, m - 1$ and $j = 1, \ldots, n - 1$. This gives a set of $(n - 1)(m - 1)$ equations in $(n + 1)(m + 1) - 4$ values of u. The value of u in each of the four corners does not appear in the central-difference analog. The boundary conditions prescribe the values for u at each of the boundary points, and the only unknown values for u in Eq. (7.55) are the interior points. This system of equations has a unique solution, which is nonzero if the boundary conditions are nonzero. It can be solved by one of the methods in Sec. 7.5.

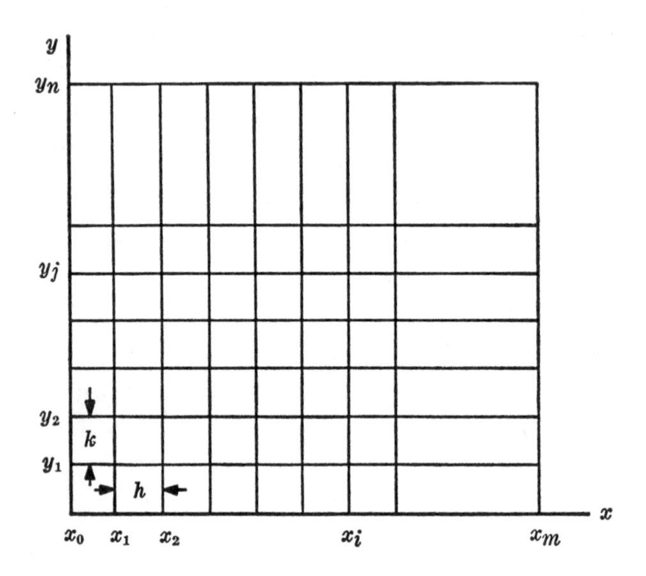

Figure 7.11 Nodal lines for a rectangular region.

In a Gauss-Seidel iteration Eq. (7.55) is solved for $u_{i,j}$.

$$u_{i,j} = \frac{u_{i+1,j} + u_{i-1,j} + r^2 u_{i,j+1} + r^2 u_{i,j-1}}{2 + 2r^2} \tag{7.56}$$

for

$$i = 1, \ldots, m - 1 \qquad j = 1, \ldots, n - 1 \qquad r^2 = \frac{h^2}{k^2}$$

A set of values is assumed for each u, and Eq. (7.56) is used for successive improvements of the approximation. A flow chart for the solution is given in Fig. 7.12. Location NØ contains a limit on the number of iterations performed, and RSQ contains r^2.

The set of simultaneous equations (7.55) can also be solved by Gaussian elimination provided a change is made in the subscript notation. The unknowns $u_{i,j}$ are converted to vector notation in order to conform with the general form for the linear algebraic

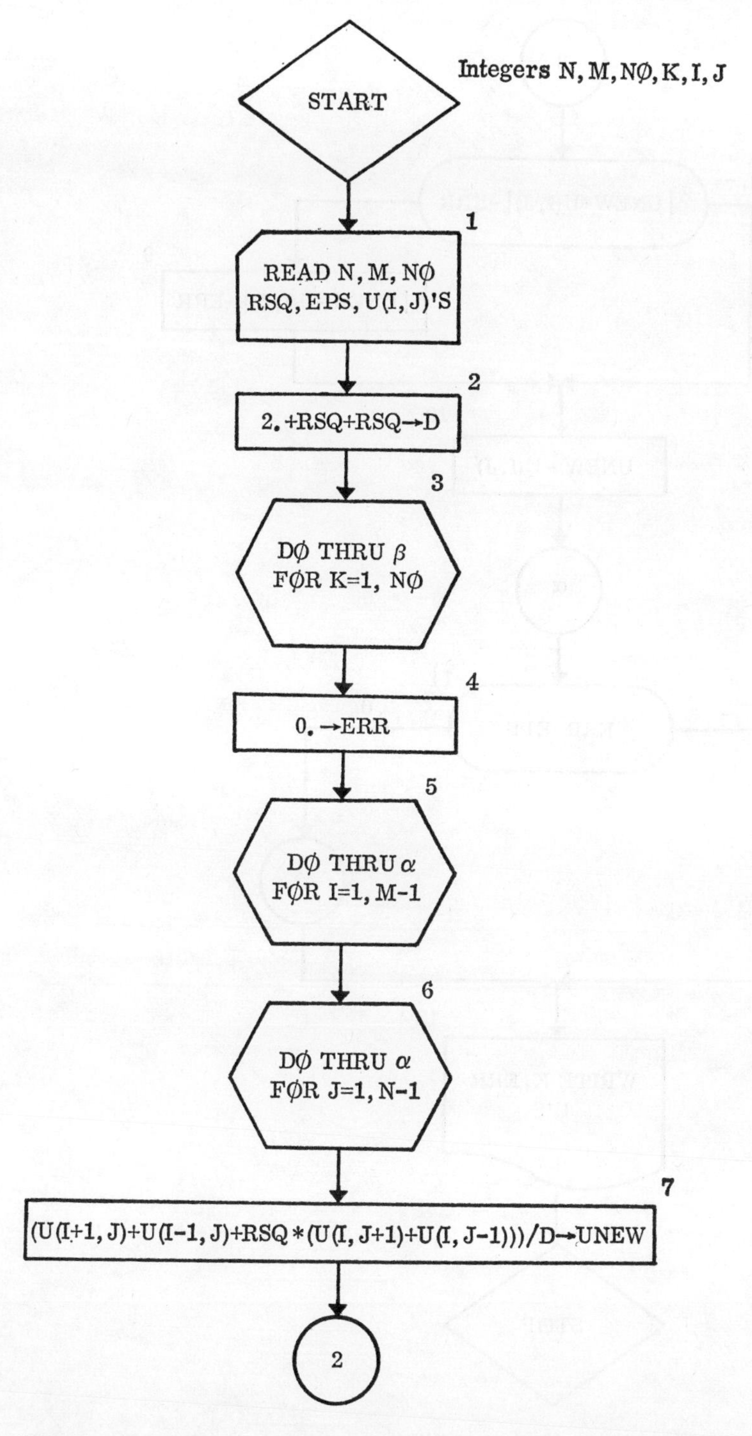

Figure 7.12 Solution of Example 7.6 by Gauss-Seidel iteration.

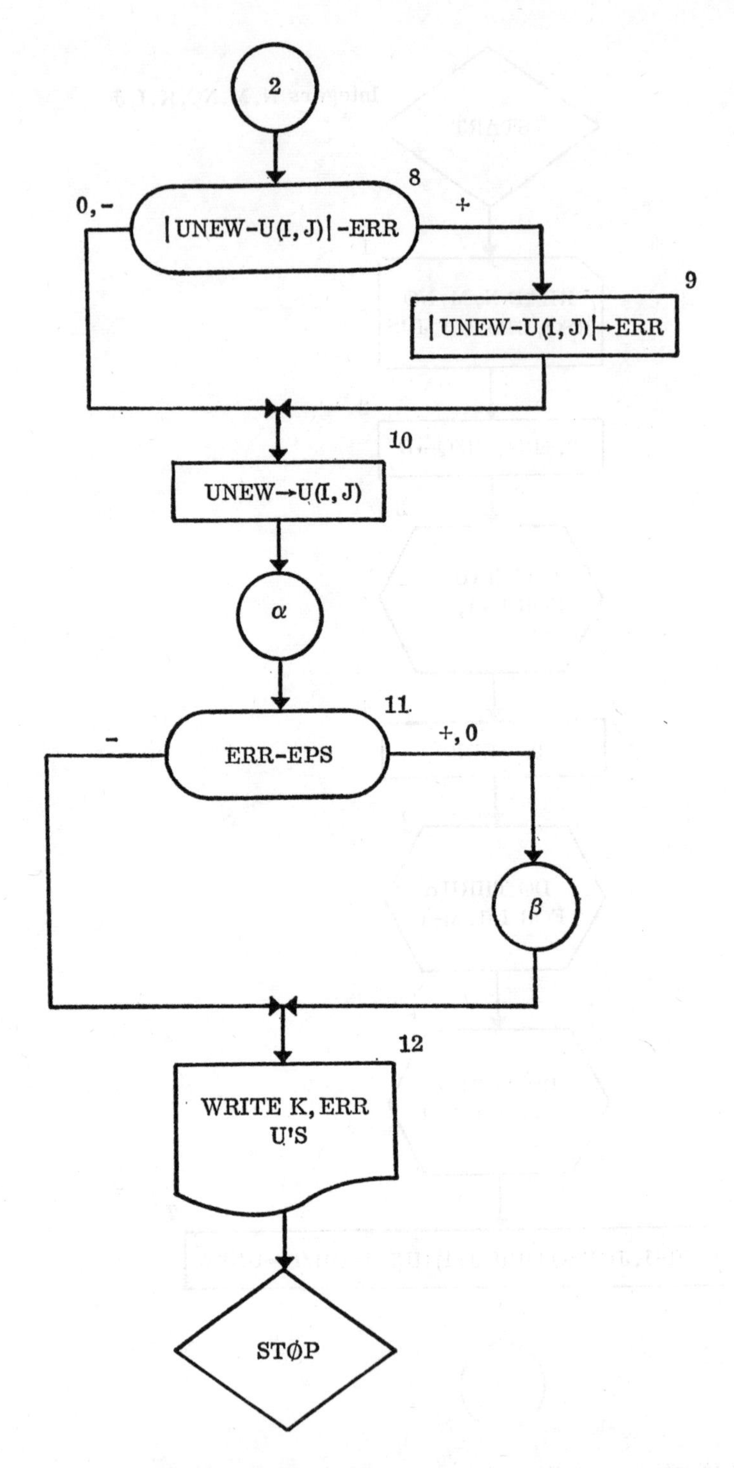

Figure 7.12 (*Continued*)

equation in (7.14). The following set of equations shows how the double-subscripted values for u can be placed into a vector V.

$$\begin{aligned} v_1 &= u_{1,1} \\ v_2 &= u_{1,2} \\ v_3 &= u_{1,3} \\ &\cdots\cdots \\ v_{m-1} &= u_{1,m-1} \\ v_m &= u_{2,1} \\ &\cdots\cdots \\ v_{(i-1)(m-1)+j} &= u_{i,j} \\ &\cdots\cdots\cdots\cdots \\ v_{(n-1)(m-1)} &= u_{n-1,m-1} \end{aligned} \tag{7.57}$$

By using the above definition for the vector V, Eq. (7.55) can be written in the form of Eq. (7.45).

$$v_{l-m+1} + r^2 v_{l-1} - 2(1 + r^2)v_l + r^2 v_{l+1} + v_{l+m-1} = b_l$$
$$l = 1, 2, \ldots, (m-1)(n-1) \tag{7.58}$$

In the transformed equation (7.58) the diagonal element of the lth equation is v_l and corresponds to $u_{i,j}$ in the lth equation of the set in Eq. (7.55). The unknown $u_{i,j-1}$ goes into v_{l-1}, $u_{i,j+1}$ into v_{l+1}, $u_{i-1,j}$ into v_{l-m+1}, and $u_{i+1,j}$ into v_{l+m-1}. Equation (7.58) is in the form of Eq. (7.45). The bandwidth parameter is equal to $m - 1$, and the coefficients $r_{l,k}$ are given below.

$$r_{l,1} = r_{l,2m-1} = 1 \qquad r_{l,m-1} = r_{l,m+1} = r^2$$
$$r_{l,m} = -2(1 + r^2) \tag{7.59}$$

All other coefficients in the equation are zero. The right-hand side of each equation is equal to zero unless the equation involves a boundary point that is nonzero. When this occurs, a constant term is introduced into the equation and is placed in b_l.

A flow chart similar to that in Fig. 7.8 is used to solve the system of equations represented by Eqs. (7.58) and (7.59).

7.7 *Accuracy of the Finite-difference Method*

In the finite-difference analog the derivatives are replaced with finite-difference approximations, which are ordinarily truncated after the first term. For the central-difference approximations the truncation error is $O(h^2)$ or $O(k^2)$. It is not possible, except in simple cases, to find a priori the effect of the truncation error on the solution. However, it seems reasonable to expect the error to be $O(h^2)$. This premise can be checked in any one problem by comparing solutions for different spacings. For example, consider the solutions given in Table 7.2. If the error term is $O(h^2)$, then

$$y_5(x) - y_{10}(x) = 4[y_{10}(x) - y_{20}(x)] \tag{7.60}$$

should hold. The subscripts denote the number of spaces used in the solution. The solutions at $x = 0.40$ are used to test the approximation.

$$y_5(0.40) - y_{10}(0.40) = -0.0000690$$

$$4[y_{10}(0.40) - y_{20}(0.40)] = -0.0000708$$

While not equal, the two numbers are close enough to support the contention that the error in the solution is $O(h^2)$.

On the basis of the above argument an improvement in the solution can be made by simply reducing the spacing in the region. However, the factor of cost enters strongly into a choice of the spacing size. This is particularly true for partial differential equations. As the spacing is reduced, the number of equations increases, and more time is required to set up and solve the finite-difference analog. In the interest of economy the spacing should be just small enough to give a suitable approximation for the solution. A test of an approximation can be made by reducing the spacing size and noting the change in the results, as suggested in Eqs. (7.14*b*). If the change is small for a significant change in the spacing, the approximation can be assumed to be satisfactory.

Results from the finite-difference method can be greatly in error when the coefficient matrix is almost singular. This condition is identified by evaluating the determinant of the coefficient matrix or in the case of Gaussian elimination by a test on the size of the divisor in each elimination step.

7.8 Conclusion

The finite-difference method proposed in this chapter is but one of the many possible methods that may be used to solve boundary-value problems. In its concept and application it is one of the simplest methods available, but the user of finite differences should always be cautious with the results. Tests with known solutions and tests using different values of a parameter can often indicate whether the method is giving satisfactory results, but this is not always possible. In some problems the number of equations required for a suitable approximation may be prohibitive.

7.9 Summary

The finite-difference method for the numerical solution of boundary-value problems is presented. The central-difference approximation for derivatives is used throughout. The method is applied to both ordinary and partial differential equations of linear form. The difference analog is developed, and several flow charts are presented for its solution.

Problems

1. Find the central-difference analog for the following boundary-value problems:

(a) $y'' + xy = c \qquad 0 < x < 1$
$y(0) = y(1) = 0$

(b) $y'' + ay' + by = cx \qquad 0 < x < l$
$y(0) = y_0 \qquad y(l) = y_l$

(c) $y'' + k^2y = x \qquad 0 < x < 1$
$y(0) = y(1) = c$

(d) $y'' + f(x)y = g^2(x) \qquad 0 < x < l$
$y(0) = a \qquad y(l) = b$

2. Find the central-difference analog for the following boundary-value problems.

(a) $y'' + g(x)y' = f(x) \qquad 0 < x < l$
$y'(0) = y(l) = 0$

(b) $y''' + f(x)y'' = g(x) \qquad 0 < x < l$
$y''(0) = y(0) = 0 \qquad y''(l) = c$

(c) $y^{IV} + p^2y'' + k^4y = g(x) \qquad 0 < x < l$
$y(0) = y''(0) = 0 \qquad y(l) = d \qquad y''(l) = c$

(d) $y^{IV} + k^4y = g(x) \qquad 0 < x < l$
$y(0) = y'(0) = 0 \qquad y''(l) = y'''(l) = 0$

Write a flow chart to compute all the entries in the matrix and store them in the proper locations for the solution of the difference analog for the following:

3. Prob. 1a **4.** Prob. 1b **5.** Prob. 1c
6. Prob. 1d **7.** Prob. 2a **8.** Prob. 2b
9. Prob. 2c **10.** Prob. 2d

Write a flow chart that will solve the matrices that are generated in the following problems:

11. Prob. 3 **12.** Prob. 4 **13.** Prob. 8
14. Prob. 9 **15.** Prob. 10

16. Write a flow chart to generate the matrix for the difference analog of Prob. 2a when a forward-difference approximation to the boundary condition at $x = 0$ is made. [Use points at $x = 0$, h, $2h$ to find the $O(h^2)$ forward-difference approximation to the first derivative at $x = 0$.]

17. Write a flow chart to generate and solve the matrix for the difference analog of Prob. 2d when a backward-difference approximation to the third derivative is made at $x = l$. [The $O(h^2)$ backward-difference approximation to the third derivative of $y(x)$ at $x = x_n$ would use the value of y at $x = x_n$, $x_n - h$, $x_n - 2h$, $x_n - 3h$, $x_n - 4h$.]

18. Write a flow chart to generate and solve the central-difference analog for the following boundary-value problem. (Use only nonzero entries in matrix.)

$$y^{IV} + f_1(x)y'' + f_2(x)y = g(x) \qquad 0 < x < l$$

$$y(0) = y''(0) = 0 \qquad y(l) = 0 \qquad y''(l) = c$$

19. The steady-state temperature distribution in a rectangular region is governed by Laplace's equation

$$\nabla^2 T = 0$$

The rectangular region is bounded by the lines $x = 0$, a and $y = 0$, b. The temperature on all boundaries except $x = a$ is 72°F, and on $x = a$ the temperature is 150°F. Write the finite-difference analog for the problem.

20. Write a flow chart to solve Prob. 19 by using the Gauss-Seidel iteration method.

21. Write a flow chart to generate the matrix and solve Prob. 19 by Gaussian elimination. (Use only the nonzero entries in the matrix.)

22. The differential equation governing the behavior of a thin plate subjected to lateral loading is

$$\nabla^2\nabla^2 w = \frac{\partial^4 w}{\partial x^4} + 2\frac{\partial^4 w}{\partial x^2\,\partial y^2} + \frac{\partial^4 w}{\partial y^4} = p(x)$$

for $0 < x < a$, $0 < y < b$. The boundary conditions are

$$w = 0 \qquad \frac{\partial^2 w}{\partial x^2} = 0 \qquad \text{at } x = 0, a$$

$$w = 0 \qquad \frac{\partial^2 w}{\partial y^2} = 0 \qquad \text{at } y = 0, b$$

Find the central-difference analog for this problem.

23. Write a flow chart to generate the matrix for Prob. 22.

24. Write a flow chart to solve Prob. 22, by Gauss-Seidel iteration.

25. Write a flow chart to solve Prob. 22, by Gaussian elimination. Use only the nonzero coefficients in the matrix.

26. Derive the central-difference analog for the following eigenvalue problems:

(a) $y'' + ky = 0 \qquad 0 < x < l$
$y(0) = y(l) = 0$

(b) $y^{\text{IV}} + k^4 y = 0 \qquad 0 < x < l$
$y(0) = y''(0) = y(l) = y''(l) = 0$

(c) $y^{\text{IV}} + p^2 y'' + k^4 y = 0 \qquad 0 < x < l$
$y(0) = y'(0) \qquad y''(l) = y'''(l) = 0$

(d) $y^{\text{VI}} + p^2 y^{\text{IV}} + k^6 y = 0 \qquad 0 < x < l$
$y(0) = y'''(0) = y^{\text{V}}(0) = 0$
$y(l) = y''(l) = 0 \qquad y^{\text{IV}}(l) = c$

27. Write a flow chart to find a value for k (an eigenvalue) in Prob. 26c that makes the characteristic determinant zero. Use the secant method and make use of the sparse matrix.

28. Write a flow chart to find the smallest eigenvalue for Prob. 26b by the power method of Chap. 4 and make use of the sparse matrix.

References

1. Forsythe, G. E., and W. R. Wasow: "Finite Difference Methods for Partial Differential Equations," John Wiley & Sons, Inc., New York, 1960.
2. Fox, L.: "Numerical Solution of Ordinary and Partial Differential Equations," Addison-Wesley Publishing Company, Inc., Reading, Mass., 1962.
3. Kunz, K. S.: "Numerical Analysis," McGraw-Hill Book Company, New York, 1957.
4. Salvadori, M. G., and M. L. Baron: "Numerical Methods in Engineering," Prentice-Hall, Inc., Englewood Cliffs, N.J., 1952.
5. Scarborough, J. B.: "Numerical Mathematical Analysis," 5th ed., The Johns Hopkins Press, Baltimore, 1962.
6. Shaw, F. S.: "Introduction to Relaxation Methods," Dover Publications, Inc., New York, 1953.
7. Southwell, R. V.: "Relaxation Methods in Engineering Science," Oxford University Press, Fair Lawn, N.J., 1940.
8. ———: "Relaxation Methods in Theoretical Physics," Oxford University Press, Fair Lawn, N.J., 1946.
9. Stanton, R. G.: "Numerical Methods for Science and Engineering," Prentice-Hall, Inc., Englewood Cliffs, N.J., 1961.
10. Young, David: The Numerical Solution of Elliptic and Parabolic Partial Differential Equations, in "Survey of Numerical Analysis," edited by John Todd, McGraw-Hill Book Company, New York, 1962.

Chapter 8

Data Approximation

8.1 Introduction This chapter presents some elementary methods for handling data records. The data may be given as a series of discrete values or as a continuous record. Examples of the former are data in tabular form, such as daily mean temperatures, yearly records of rainfall, or tables of trigonometric functions. Recording transducers such as accelerometers, thermometers, and pressure gages may be set up to give a continuous output, in which case the data form a continuous record. The chapter is concerned primarily with the problem of approximating a data record with some functional relationship between the variables. To facilitate future discussion some elementary statistical terms will first be defined.

The data to be handled are a collection of numbers $y_1, y_2, \ldots, y_m$, which may represent the ordinates of a trace for corresponding abscissas $x_1, x_2, \ldots, x_m$. The ordinates will be represented by the vector Y and the abscissas by the vector X.

The mean value of Y is the sum of the numbers $y_1, \ldots, y_m$ divided by m. The mean is designated by the expression Y_{avg} or μ_Y.

$$\mu_Y = Y_{\text{avg}} = \frac{1}{m} \sum_{j=1}^{m} y_j \tag{8.1}$$

The mean (or average) of Y is a measure of the expected value for Y. The subscript on μ denotes the collection to which it refers.

The variance of Y is a measure of the departure of Y from its mean value and is defined by

$$Y_{\text{var}} = \frac{1}{m} \sum_{j=1}^{m} (y_j - \mu_Y)^2 \tag{8.2}$$

Each term in the sum is a positive number, so that any deviation of a value of y_j from the mean will contribute a positive number to the sum. Thus Y_{var} is a positive number.

The standard deviation of the collection Y is defined as the positive square root of the variance. The standard deviation will be denoted by σ with a subscript to denote which collection is used:

$$\sigma_Y = \sqrt{Y_{\text{var}}} = \sqrt{\frac{1}{m} \sum_{j=1}^{m} (y_j - \mu_Y)^2} \tag{8.3}$$

Since Y_{var} is the standard deviation squared, it is often written as $\sigma_Y{}^2$.

8.2 Data Problem in Two Variables

Let a set of values Y be defined for corresponding values of X, i.e., for each x_j there is a corresponding y_j. Such a set of values may be any information that can be represented graphically in a two-dimensional plot. Examples are plots of force versus displacements,

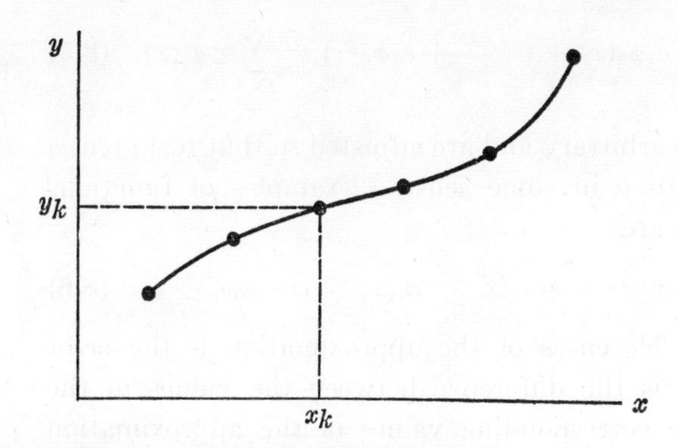

Figure 8.1 Data curve.

position versus time, rainfall versus temperature, or gas mileage versus speed. In these cases the values Y are the set of numerical values of the ordinate at discrete values of the abscissa (Fig. 8.1). The question arises: How is Y related to X? How can a functional relationship be found? And how can it be tested for validity? Some elementary techniques are presented for finding a mathematical approximation for the relationship between two variables and then testing the approximation to determine how well it represents the actual data.

8.3 An Approximate Function to a Data Record

There are several criteria that can be used as a guide in establishing a relationship between X and Y. It may be desirable to keep the relationship simple, in which case a linear function can be used:

$$y = c_1 + c_2 x \tag{8.4}$$

Or if a closer fit is needed, a more complex relationship between x and y will be necessary. The problem is formulated as one of approximating y as a function of x on an interval $a \leq x \leq b$ to give the best fit for the available data. Let it be assumed that the approximating function $f(x)$ is a finite sum of known functions of x. Let these known functions of x be denoted by $\phi_i(x)$, where i goes from 1 to n:

$$f(x) = c_1\phi_1(x) + c_2\phi_2(x) + \cdots + c_n\phi_n(x) = \sum_{i=1}^{n} c_i\phi_i(x) \tag{8.5}$$

The coefficients c_i are arbitrary and are adjusted so that $f(x)$ gives a best approximation to y in some sense. Examples of functions $\phi_i(x)$ commonly used are:

$$\phi_i(x) = x^{i-1} \qquad \phi_i(x) = \sin ix \qquad \phi_i(x) = \cos w_i x \tag{8.6}$$

The measure of closeness of the approximation is the error function $e(x)$, which is the difference between the values in the collection Y and the corresponding values of the approximation $f(x)$. Thus the value of the error function at $x = x_j$ is

$$e(x_j) = y_j - f(x_j) \tag{8.7}$$

The data may be such that y is defined for every value of x on some interval $a \leq x \leq b$, or y may be defined only at a finite number of discrete values of x. In the continuous (first) case the error function is defined at every value of x, whereas in the discrete (second) case it is defined only at the given $x = x_j$. In both cases the criterion for choosing values for the coefficients $c_1, c_2, \ldots, c_n$ is to make the error function small.

There are different criteria for making the error function small, e.g., assuring zero error at p selected values for x. This will require p coefficients in the function $f(x)$ and will lead to p simultaneous linear algebraic equations for the coefficients c_i, $i = 1, 2, \ldots, p$. This is sometimes referred to as the *method of collocation*. Another criterion for finding the coefficients in the approximating function is to minimize the largest error on the interval of approximation, i.e., to make the largest of the error terms given by Eq. (8.7) a minimum. This is the *minimax*, or Chebyshev, *approximation*. A common technique for determining the coefficients is to make the sum of the squares of the errors given by Eq. (8.7) a minimum. This is the *method of least squares*. The collocation and least-squares methods are discussed in the following sections. The minimax method is considerably more complex, and is not taken up in this text. First, however, the problem is put into the more convenient matrix notation.

At the set of points $x = x_j$, $j = 1, 2, \ldots, m$, the equation for the error, which is given by Eq. (8.7), is written in terms of the coefficients c_i, $i = 1, 2, \ldots, n$, as

$$e_j = y_j - f(x_j) = y_j - \sum_{i=1}^{n} c_i\phi_i(x_j) \qquad j = 1, 2, \ldots, m \tag{8.8}$$

This is a system of m linear equations in the n coefficients c_i. The equations can be written in matrix form when the following definitions are used.

1. Vectors E and Y are m-dimensional column vectors with elements e_j and y_j, respectively.
2. The elements c_i form the n-dimensional column vector C.
3. The elements $\phi_j(x_i)$, $j = 1, 2, \ldots, n$, $i = 1, 2, \ldots, m$, form the entries $a_{i,j}$ in the matrix A.

$$C = \begin{bmatrix} c_1 \\ c_2 \\ \cdot \\ \cdot \\ \cdot \\ c_n \end{bmatrix} \qquad E = \begin{bmatrix} e_1 \\ e_2 \\ \cdot \\ \cdot \\ \cdot \\ e_m \end{bmatrix} \qquad Y = \begin{bmatrix} y_1 \\ y_2 \\ \cdot \\ \cdot \\ \cdot \\ y_m \end{bmatrix}$$

$$A = \begin{bmatrix} \phi_1(x_1) & \phi_2(x_1) & \cdots & \phi_n(x_1) \\ \phi_1(x_2) & \phi_2(x_2) & \cdots & \phi_n(x_2) \\ \cdots & \cdots & \cdots & \cdots \\ \phi_1(x_m) & \phi_2(x_m) & \cdots & \phi_n(x_m) \end{bmatrix} \tag{8.9}$$

With these definitions Eq. (8.8) can be represented by the matrix equation

$$E = Y - AC \tag{8.10}$$

8.4 Method of Collocation

In the method of collocation the approximate function $f(x)$ passes through a specified number of points in the collection. Let the points through which $f(x)$ is to pass be designated (x_i,y_i), $i = 1, 2, \ldots, n$. The approximate function in Eq. (8.5) has n terms, and there will be n unknown values for c, which are chosen so that the difference between $f(x)$ and the data is equal to zero at the specified points (x_i,y_i), $i = 1, 2, \ldots, n$. The error vector in Eq. (8.10) is identically zero, and the equation reduces to a system of n linear equations in n unknowns.

$$AC = Y \tag{8.11}$$

A is the n by n matrix obtained from Eq. (8.9), where the number of functions ϕ_i is equal to the number of points through which the function is to pass. C and Y are as defined in Eq. (8.9) with $m = n$.

The approximate function $f(x)$ obtained from the collocation method fits the data exactly at the selected points, but the error for other values of x may become large. The method of least squares, given below, offers a technique for keeping the error between an approximate function and the data small for all points in the collection.

8.5 Least-squares Approximation

The error between the given data and the approximating function at x_j is given by

$$y_j - f(x_j) = e(x_j) \tag{8.8}$$

If this error is squared and then summed on j, a positive number, which is designated γ^2, is obtained, and

$$\gamma^2 = \sum_{j=1}^{m} [e(x_j)]^2 \tag{8.12}$$

where m is the number of data points. If y is a continuous function of x, then γ^2 is defined as an integral

$$\gamma^2 = \int_a^b [e(x)]^2\, dx \tag{8.13}$$

In either the continuous or discrete case the object've in the method of least squares is to choose the constants in $f(x)$ so that γ^2 will have a minimum value.

Consider first the problem where y is defined only at discrete values for $x = x_1, x_2, \ldots, x_m$. The function γ^2 can be written in terms of the error vector E:

$$\gamma^2 = \sum_{j=1}^{m} [e(x_j)]^2 = E^tE \tag{8.14}$$

where E^t is the transpose of E. The problem is posed as one of minimizing γ^2, which is a function of the coefficient vector C:

$$\gamma^2 = \gamma^2(c_1, c_2, \ldots, c_n) = \gamma^2(C) \tag{8.15}$$

Let C_0 be the coefficient vector such that $\gamma^2(C_0)$ is a minimum value, and let C be the vector defined by

$$C = C_0 + sP \tag{8.16}$$

where s is an arbitrary scalar and P is an arbitrary nonzero vector. The vectors C_0 and P will be regarded as fixed, and the error vector E is then a function of the scalar s:

$$\begin{aligned} E(s) &= Y - AC = Y - A(C_0 + sP) \\ &= Y - AC_0 - sAP = E_0 - sQ \end{aligned} \tag{8.17}$$

where $E_0 = Y - AC_0$ and $Q = AP$. The function γ^2 will also be a function of s:

$$\begin{aligned} \gamma^2(s) &= E(s)^tE(s) \\ &= (E_0{}^t - sQ^t)(E_0 - sQ) \\ &= E_0{}^tE_0 - s(Q^tE_0 + E_0{}^tQ) + s^2(Q^tQ) \end{aligned} \tag{8.18}$$

Using the fact that $E_0{}^tQ = Q^tE_0 = P^tA^tE_0$ and letting $E_0{}^tE_0 = \gamma_0{}^2$ results in

$$\gamma^2(t) = \gamma_0{}^2 - 2P^t(A^tE_0)s + (Q^tQ)s^2 \tag{8.19}$$

where Q^tQ is the sum of the elements of Q squared and hence is a positive number. Thus Eq. (8.19) represents a parabola in the γ^2, s plane opening in the positive γ^2 direction.

If the derivative of γ^2 with respect to s at $s = 0$ is not zero, $\gamma^2(s)$ will take on a value less than $\gamma_0{}^2$ for some value of $s \neq 0$. This is contrary to the assumption that C_0 is the vector that minimizes γ^2; hence the derivative of γ^2 with respect to s must be zero at $s = 0$.

$$\frac{d\gamma^2(0)}{ds} = -2P^t(A^tE_0) = 0 \tag{8.20}$$

Equation (8.20) holds for all choices of the vector P. If P is set equal to A^tE_0, Eq. (8.20) represents the sum of the elements of P squared. If this sum is to be zero, each element of P must be zero, or

$$A^tE_0 = A^t(Y - AC_0) = 0 \tag{8.21}$$

The vector C_0 must satisfy Eq. (8.21) in order for $\gamma^2(C_0)$ to be a minimum. Since Eq. (8.19) is a parabola opening upward, Eq. (8.21) is also a sufficient condition for $\gamma^2(C_0)$ to be a minimum. Equation (8.21) is written

$$BC_0 = V \tag{8.22}$$

where

$$B = A^tA \qquad V = A^tY \tag{8.23}$$

B is an n by n matrix, and V is an n-dimensional vector. The elements of B and V can be evaluated when the elements of A and Y in Eq. (8.9) are known. The elements of B and V are given in terms of the elements of A and Y.

$$b_{ij} = \sum_{k=1}^{m} a_{ki}a_{kj} \qquad v_i = \sum_{k=1}^{m} a_{ki}y_k \qquad \text{for } i, j = 1, 2, \ldots, n \tag{8.24}$$

Once the elements of B and V are computed, Eq. (8.22) can be solved for C_0 by using one of the methods described in Chap. 4. The equations represented by (8.22) are called the *least-squares* or *normal equations*.

For the function $f(x)$ given in

$$f(x) = \sum_{i=1}^{n} c_i\phi_i(x) \tag{8.5}$$

the elements of the matrix A are given by

$$a_{ki} = \phi_i(x_k) \tag{8.9}$$

The elements of B and V are

$$\left.\begin{aligned} b_{ij} &= \sum_{k=1}^{m} \phi_i(x_k)\phi_j(x_k) \\ v_i &= \sum_{k=1}^{m} \phi_i(x_k)y_k \end{aligned}\right\} \quad \text{for } i, j = 1, 2, \ldots, n \tag{8.25}$$

One and only one solution for C_0 exists if the functions $\phi_i(x)$ are linearly independent. This means that a set of nonzero coefficients $d_1, d_2, \ldots, d_n$ that will satisfy Eq. (8.26) for any x cannot exist.

$$d_1\phi_1(x) + d_2\phi_2(x) + \cdots + d_n\phi_n(x) = 0 \tag{8.26}$$

The value of γ^2 will provide a measure of the closeness of the approximation to the original curve. The following expression for γ^2 can be readily evaluated provided the elements of V are retained:

$$\gamma^2 = \sum_{k=1}^{m} y_k^2 - \sum_{i=1}^{n} c_i v_i \tag{8.27}$$

The terms $c_1, c_2, \ldots, c_n$ are the elements of the vector C_0. The proof of this formula is not given. It follows from the definition of γ^2 and the condition that the values of $c_1, c_2, \ldots, c_n$ are found from the least-squares equations.

Some example problems are solved to illustrate the method of least squares in finding approximate functions to a set of data.

Example 8.1

Find the linear approximation to $y = \sqrt{x}$ on the range $0.1 \leq x \leq 1.0$ by using the 10 points $x = x_k = 0.1k$, $k = 1, 2, \ldots, 10$, for the data points.

In the linear function $f(x)$ the constants c_1 and c_2 are evaluated by using the least-squares equations.

$$f(x) = c_1 + c_2 x$$

From Eq. (8.5) $\phi_1(x) = 1$ and $\phi_2(x) = x$. The coefficients b_{ij} and v_i are computed from Eqs. (8.25).

$$b_{11} = \sum_{k=1}^{10} \phi_1(x_k)\phi_1(x_k) = \sum_{k=1}^{10} 1 = 10$$

$$b_{12} = b_{21} = \sum_{k=1}^{10} \phi_1(x_k)\phi_2(x_k) = \sum_{k=1}^{10} x_k = 5.5$$

$$b_{22} = \sum_{k=1}^{10} \phi_2(x_k)\phi_2(x_k) = \sum_{k=1}^{10} x_k^2 = 3.85$$

$$v_1 = \sum_{k=1}^{10} y_k = \sum_{k=1}^{10} \sqrt{x_k} = 7.105093$$

$$v_2 = \sum_{k=1}^{10} x_k y_k = \sum_{k=1}^{10} \sqrt{x_k}^3 = 4.511605$$

Substitution of these numbers into Eq. (8.22) gives the following scalar equations in c_1 and c_2:

$$10.0c_1 + 5.5c_2 = 7.105093$$

$$5.5c_1 + 3.85c_2 = 4.511695$$

The solution of these equations is

$$c_1 = 0.307914$$

$$c_2 = 0.731992$$

The value for γ^2 is found from Eq. (8.27).

$$\gamma^2 = \sum_{k=1}^{10} y_k^2 - c_1 v_1 - c_2 v_2 = 0.009718$$

The curves $y = \sqrt{x}$ and $y = f(x)$ are sketched in Fig. 8.2. The approximation is fairly good on the range $0.1 \leq x \leq 1.0$, which is the interval of the approximation. This approximation to $\sqrt{x}$ is sometimes used to obtain a first approximation to the square root of a number in the range $0.1 \leq x \leq 1.0$. The approximation might then be improved by one of the methods in Chap. 3.

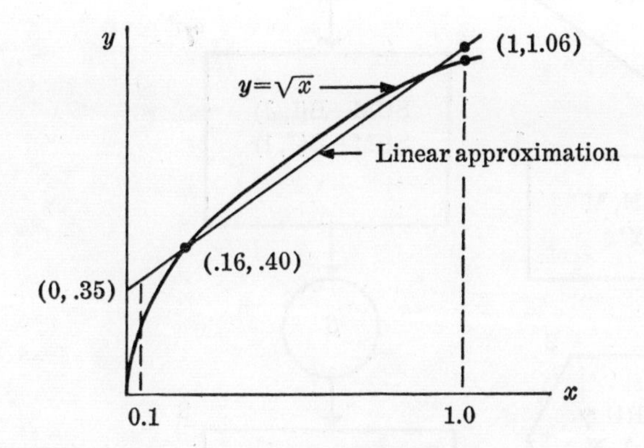

Figure 8.2 Linear approximation to $\sqrt{x}$.

A flow chart to compute the coefficient vector C_0 is given in Fig. 8.3, and a short description of it follows.

2–7. Compute $b_{i,j}$.

$$b_{i,j} = \sum_{k=1}^{m} \phi_i(x_k)\phi_j(x_k) \qquad \text{for } i, j = 1, 2, \ldots, n$$

Note that $b_{i,j} = b_{j,i}$, so that only $(n^2 + n)/2$ numbers need be computed in this operation.

8–11. Compute v_i.

$$v_i = \sum_{k=1}^{m} \phi_i(x_k)y_k \qquad \text{for } i = 1, 2, \ldots, n$$

These numbers are saved in the vector V for use in the computation of γ^2.

12. Solve the matrix using one of the procedures described in Chap. 4 and place the solution in vector C.

13–17. Compute γ^2.

$$\gamma^2 = \sum_{j=1}^{m} y_j^2 - \sum_{i=1}^{n} c_i v_i$$

This value is stored in SUM.

When the function $y(x)$ is defined for every value of x, the definition of γ^2 involves an integral instead of a sum.

$$\gamma^2 = \int_a^b [e(x)]^2\, dx \tag{8.13}$$

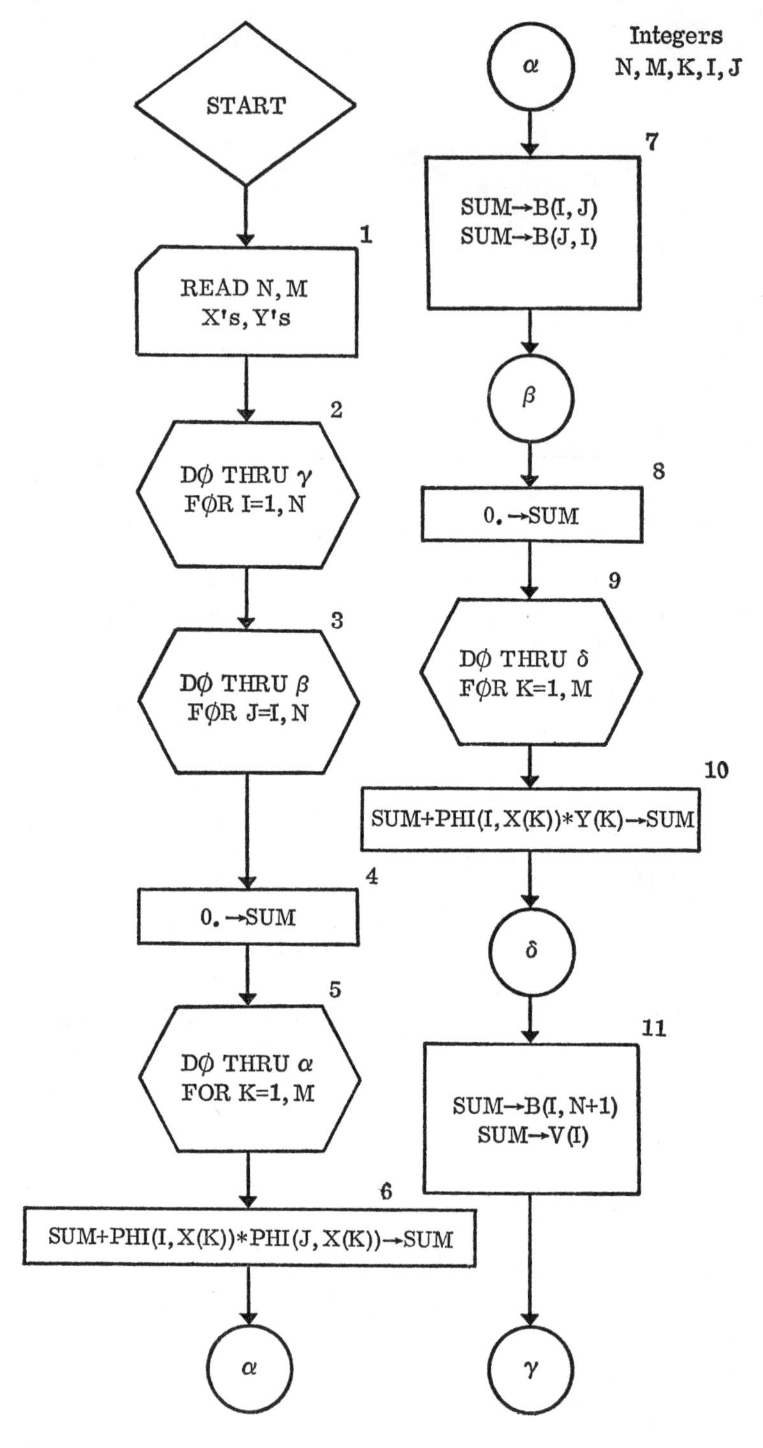

Figure 8.3 Least-squares approximation to a set of points. [*Note:* PHI(I,X) is the function $\phi_i(x)$.]

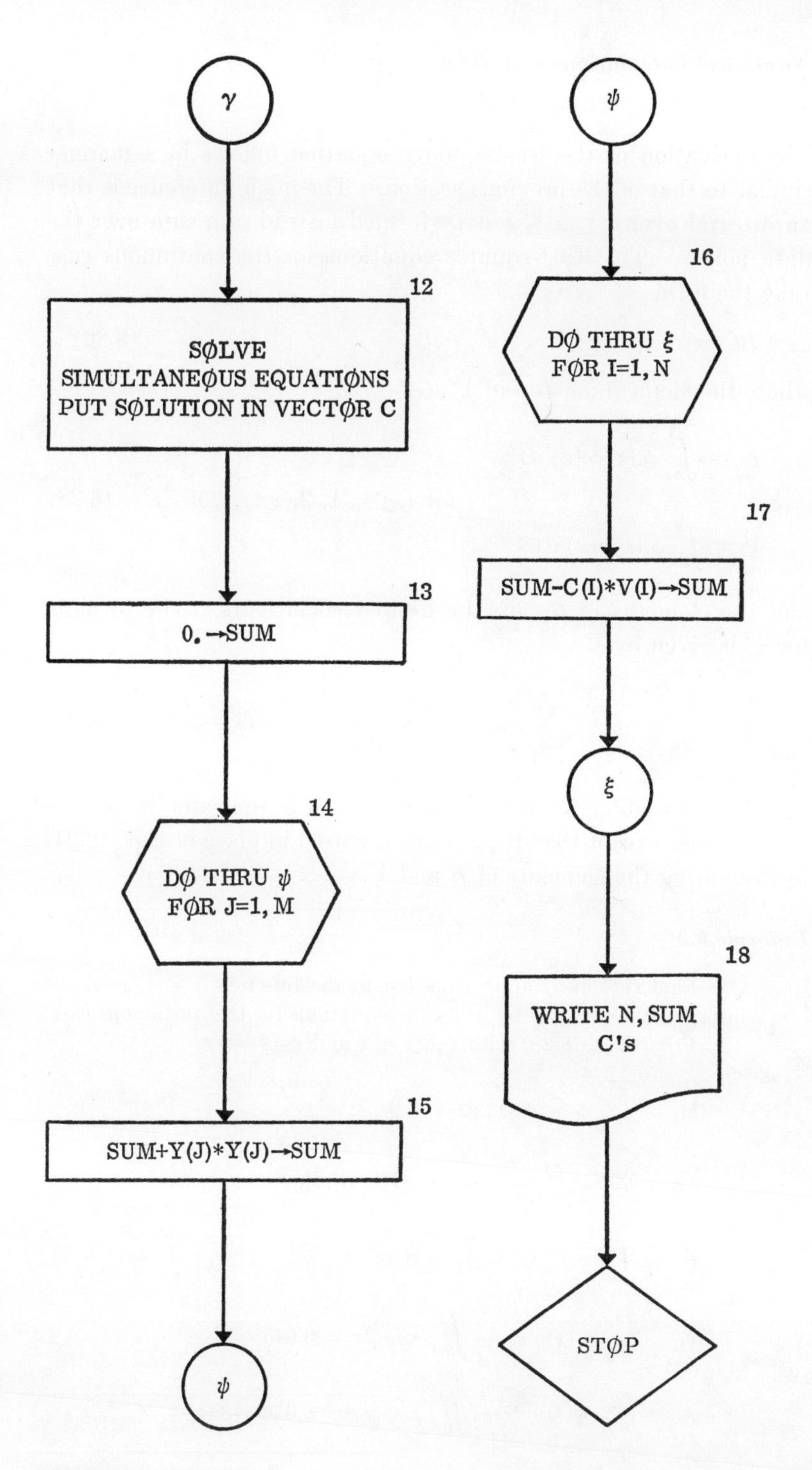

Figure 8.3 (*Continued*)

The derivation of the least-squares equation follows in a manner similar to that of the previous section. The main difference is that an integral over $a \leq x \leq b$ is performed instead of a sum over the data points. The least-squares equations for the continuous case take the form

$$BC_0 = V \tag{8.22}$$

where the elements of B and V are

$$\left.\begin{aligned} b_{ij} &= \int_a^b \phi_i(x)\phi_j(x)\,dx \\ v_i &= \int_a^b \phi_i(x)y(x)\,dx \end{aligned}\right\} \quad \text{for } i, j = 1, 2, \ldots, n \tag{8.28}$$

and the elements of C_0 are the desired coefficients. The formula for γ^2 is given by

$$\gamma^2 = \int_a^b y^2\,dx - \sum_{i=1}^{n} c_i v_i \tag{8.29}$$

The procedure for the continuous case is the same as for the discrete case except that Eqs. (8.28) are used in place of Eqs. (8.24) in evaluating the elements of B and V.

Example 8.2

Find the linear approximation to the function $y = \sqrt{x}$ on the interval $0.1 \leq x \leq 1$ by using the equations for the continuous case. From Eqs. (8.28) with $\phi_1(x) = 1$ and $\phi_2(x) = x$,

$$b_{11} = \int_{0.1}^{1} \phi_1(x)\phi_1(x)\,dx = 0.9$$

$$b_{12} = b_{21} = \int_{0.1}^{1} \phi_1(x)\phi_2(x)\,dx = 0.495$$

$$b_{22} = \int_{0.1}^{1} \phi_2(x)\phi_2(x)\,dx = 0.33$$

$$v_1 = \int_{0.1}^{1} \phi_1(x)y\,dx = \int_{0.1}^{1} \sqrt{x}\,dx = 0.6455838$$

$$v_2 = \int_{0.1}^{1} \phi_2(x)y\,dx = \int_{0.1}^{1} x\sqrt{x}\,dx = 0.3987351$$

Substitution of these numbers into Eq. (8.22) gives

$$0.900c_1 + 0.495c_2 = 0.6455838$$

$$0.495c_1 + 0.333c_2 = 0.3987351$$

The solution of these equations is

$$c_1 = 0.322003$$

$$c_2 = 0.718749$$

The value for γ^2 is computed from Eq. (8.29):

$$\gamma^2 = \int_{0.1}^{1} y^2\,dx - c_1v_1 - c_2v_2 = 0.000530$$

In this example the values of c_1 and c_2 are little different from the discrete case and the two approximating functions are very nearly equal to each other. The value of γ^2 is, however, an order of magnitude different. This would indicate that the value of γ^2 may be a poor criterion for the fit of an approximating function. This is borne out in a later discussion.

Example 8.3

Find the least-squares polynomial of degree $n - 1$ to approximate the function $y = e^x$ on the interval $0 \leq x \leq l$ by using the equations for the continuous case.

The elements of the coefficient matrix in Eq. (8.22) are found from Eqs. (8.28) by setting $b = l$, $a = 0$, and $\phi_i(x) = x^{i-1}$:

$$b_{ij} = \int_0^l x^{i+j-2}\,dx = \frac{l^{i+j-1}}{i+j-1}$$

$$v_i = \int_0^l x^{i-1}e^x\,dx \tag{8.29a}$$

For $i = 1$,

$$v_1 = e^x \Big|_0^l = e^l - 1 \tag{8.29b}$$

For $i > 1$, the integration for v_i is done by parts.

$$v_i = x^{i-1}e^x \Big|_0^l - (i - 1) \int_0^l x^{i-2}e^x\,dx \tag{8.29c}$$

By definition the integral in the expression for v_i is equal to v_{i-1}. Therefore

$$v_i = l^{i-1}e^l - (i - 1)v_{i-1} \qquad \text{for } i = 2, 3, \ldots, n$$

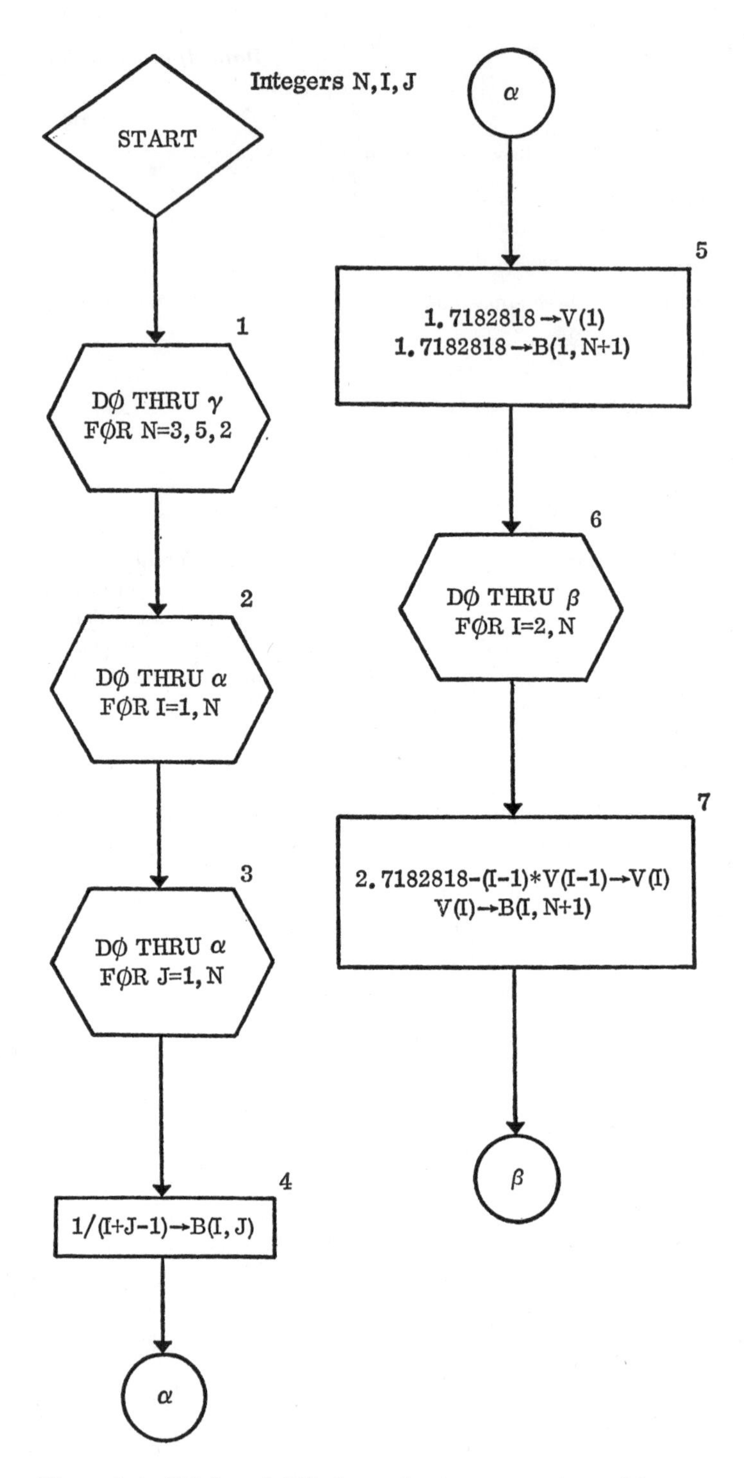

Figure 8.4 Third- and fifth-degree least-squares polynomials to fit e^x over $0 \leq x \leq 1$.

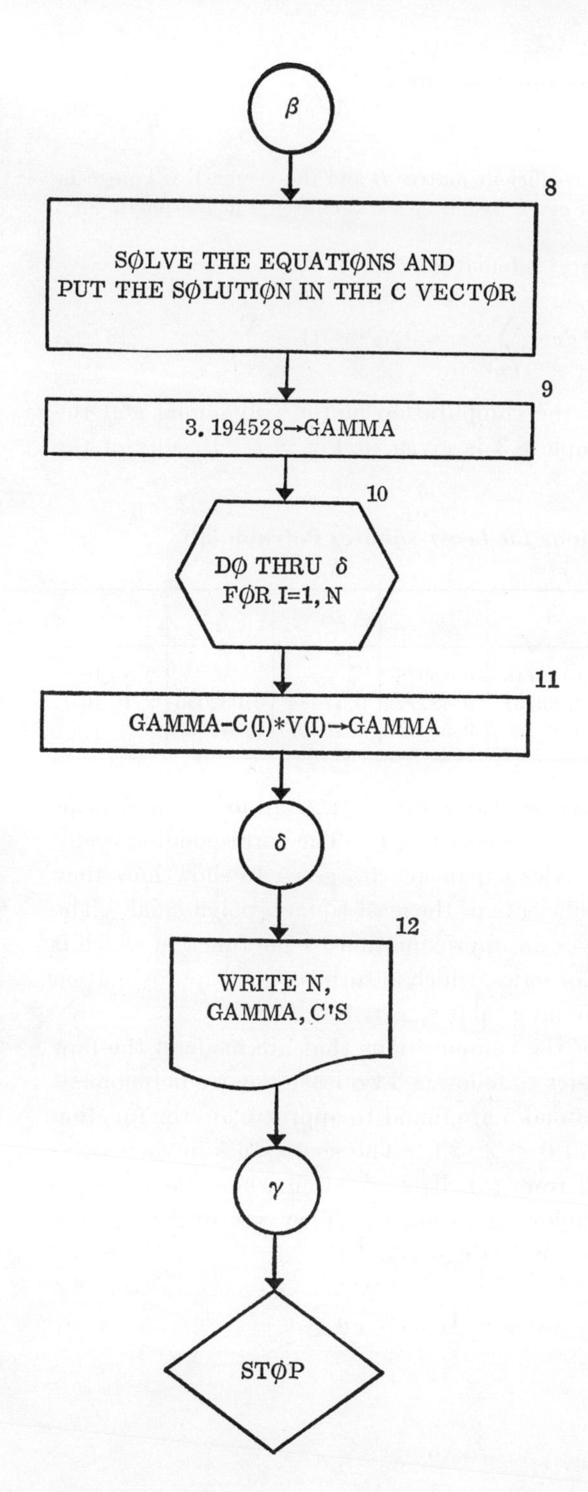

Figure 8.4 (*Continued*)

Each term in the coefficient matrix B and the vector V is known in terms of l. For a given l each of the coefficients in the polynomial can be computed.

The value for γ^2 is found from Eq. (8.29).

$$\gamma^2 = \int_0^l (e^x)^2\,dx - \sum_{i=1}^{n} c_i v_i = \tfrac{1}{2}(e^{2l} - 1) - \sum_{i=1}^{n} c_i v_i \qquad (8.29d)$$

A flow chart for the computation of the polynomial and the value of γ^2 for Example 8.3 is given in Fig. 8.4. Results of the

Table 8.1
Results of Computations for Least-squares Polynomial for e^x on $0 \leq x \leq 1$

	c_1	c_2	c_3	c_4	c_5	γ^2
$n = 3$	1.01299	0.85118	0.83918			$2.8 \cdot 10^{-5}$
$n = 5$	0.99788	1.00361	0.48772	0.17484	0.052036	$2.7 \cdot 10^{-7}$
Taylor series	1.0	1.0	0.5	0.16667	0.041667	

computations are given in Table 8.1 for a third-degree and fifth-degree polynomial when l is equal to 1. The corresponding coefficients in the Taylor-series expansion are given to show how they compare with the coefficients of the least-squares polynomial. The fifth-degree curve gives an approximation to the function which is very close to the Taylor series, which in turn is a good approximation to the function on the interval $0 \leq x \leq 1$.

A brief review of the computations that are made in the flow chart will make it easier to follow. Two least-squares polynomials of degree n equal to 3 and 5 are found to approximate the function $y = e^x$ on the interval $0 \leq x \leq 1$. The coefficients in each polynomial are computed from a matrix equation whose elements are determined by the following formulas. They are obtained from Eqs. (8.29a) to (8.29c) by letting l equal 1.

$$b_{i,j} = \frac{1}{i + j - 1} \qquad i, j = 1, \ldots, n$$

$$v_1 = e - 1$$

$$v_i = e - (i - 1)v_{i-1} \qquad i = 2, \ldots, n$$

Additionally, it is required to find the value of γ^2, where

$$\gamma^2 = \tfrac{1}{2}(e^2 - 1) - \sum_{i=1}^{n} c_i v_i$$

8.6 The Least-squares Polynomial

The set of functions most commonly used in the least-squares approximation to a set of points is

$$\phi_i(x) = x^{i-1} \qquad i = 1, 2, \ldots, n \tag{8.30}$$

Substituting this function into $f(x)$ in Eq. (8.5) gives the following $(n - 1)$st-degree polynomial in x:

$$f(x) = c_1 + c_2 x + c_3 x^2 + \cdots + c_n x^{n-1} \tag{8.31}$$

When the coefficients $c_1, c_2, \ldots, c_n$ are evaluated for a best fit in the least-squares sense, this polynomial is called the *least-squares polynomial.* The elements of the B and V matrices in Eq. (8.22) are obtained from Eqs. (8.25) or (8.28), as has been illustrated in the examples.

When y is a continuous function of x, the matrix coefficients are obtained from Eqs. (8.28) as follows:

$$b_{ij} = \int_a^b \phi_i(x)\phi_j(x)\,dx = \int_a^b x^{i-1}x^{j-1}\,dx = \int_a^b x^{i+j-2}\,dx$$

$$= \left[\frac{x^{i+j-1}}{i+j-1}\right]_a^b = \frac{b^{i+j-1} - a^{i+j-1}}{i+j-1} \qquad \text{for } i, j = 1, 2, \ldots, n \tag{8.32}$$

and

$$v_i = \int_a^b \phi_i(x) y\,dx = \int_a^b x^{i-1} y\,dx \quad \text{for } i = 1, 2, \ldots, n \tag{8.33}$$

When y is a discrete function of x, the matrix coefficients are obtained from Eqs. (8.25) and are given by

$$b_{ij} = \sum_{k=1}^{m} \phi_i(x_k)\phi_j(x_k) = \sum_{k=1}^{m} x_k^{i+j-2} \qquad \text{for } i, j = 1, 2, \ldots, n \tag{8.34}$$

and

$$v_i = \sum_{k=1}^{m} \phi_i(x_k) y_k = \sum_{k=1}^{m} x_k^{i-1} y_k \qquad \text{for } i = 1, 2, \ldots, n \tag{8.35}$$

Note in both the continuous and the discrete cases that $b_{ij} = b_{pr}$ provided that $i + j = p + r$. This fact is used in the computations by defining a vector S as follows. For the continuous case

$$b_{ij} = s_p = \frac{b^p - a^p}{p} \qquad \text{for } p = i + j - 1 = 1, 2, \ldots, 2n - 1 \quad (8.36)$$

For the discrete case

$$b_{ij} = s_p = \sum_{k=1}^{m} x_k^{p-1} \qquad \text{for } p = i + j - 1 = 1, 2, \ldots, 2n - 1 \quad (8.37)$$

The use of Eqs. (8.36) and (8.37) will greatly reduce the number of computations for evaluating the elements of B. The solution for a least-squares polynomial of degree $n - 1$ is illustrated with Example 8.4.

Example 8.4

Find the least-squares polynomial of degree $n - 1$ for the set of points x_k, y_k, $k = 1, 2, \ldots, m$, and find the value for γ^2. The elements of the matrix for finding the coefficients in the least-squares polynomial are computed by using Eqs. (8.35) and (8.37).

$$s_p = \sum_{k=1}^{m} x_k^{p-1} \qquad p = 1, 2, \ldots, 2n - 1$$

$$v_i = \sum_{k=1}^{m} x_k^{i-1} y_k \qquad i = 1, 2, \ldots, n$$

Equation (8.38) shows how these elements are arranged in the matrix.

$$\begin{bmatrix} s_1 & s_2 & s_3 & \cdots & s_n & v_1 \\ s_2 & s_3 & s_4 & \cdots & s_{n+1} & v_2 \\ \cdot & \cdot & \cdot & \cdots & \cdot & \cdot \\ s_n & s_{n+1} & s_{n+2} & \cdots & s_{2n-1} & v_n \end{bmatrix} \quad (8.38)$$

Before this matrix can be solved for the coefficients in the least-squares polynomial, the elements in S must be transferred to the matrix B. This transfer is accomplished by setting

$$b_{i,j} = s_{i+j-1} \qquad \text{for } i, j = 1, 2, \ldots, n$$

The computational procedure for finding the least-squares polynomial and the value of γ^2 is outlined below:

1. Compute the elements of the vector S

$$s_i = \sum_{k=1}^{m} x_k^{i-1} \qquad i = 1, 2, \ldots, 2n - 1$$

2. Compute the elements of the vector V

$$v_i = \sum_{k=1}^{m} x_k^{i-1} y_k \qquad i = 1, 2, \ldots, n$$

3. Transfer S and V to the appropriate locations in B.

$$b_{i,j} = s_{i+j-1}$$

$$b_{i,n+1} = v_i$$

4. Solve the B matrix for the coefficients c_i, $i = 1, \ldots, n$, in the least-squares polynomial.

5. Compute γ^2.

$$\gamma^2 = \sum_{k=1}^{m} y_k^2 - \sum_{i=1}^{n} c_i v_i$$

A flow chart constructed along the lines of this procedure is shown in Fig. 8.5.

8.7 Some Comments on Least-squares Polynomials

Care should be taken in choosing the degree of polynomial to use when approximating a function with a least-squares polynomial. If the degree is too small, the fit can be poor. On the other hand, if the degree is large, the fit may be poor near the end points of the interval. It can in fact be shown by using Legendre polynomials that a polynomial approximation will not converge at the end points of the interval of approximation regardless of the number of terms used.[4]

In practice, when trying to approximate a function by a polynomial, a value is chosen for n, and the least-squares polynomial of degree n is found. Then if this approximation is not satisfactory, n is increased, and the procedure is repeated. However, there is

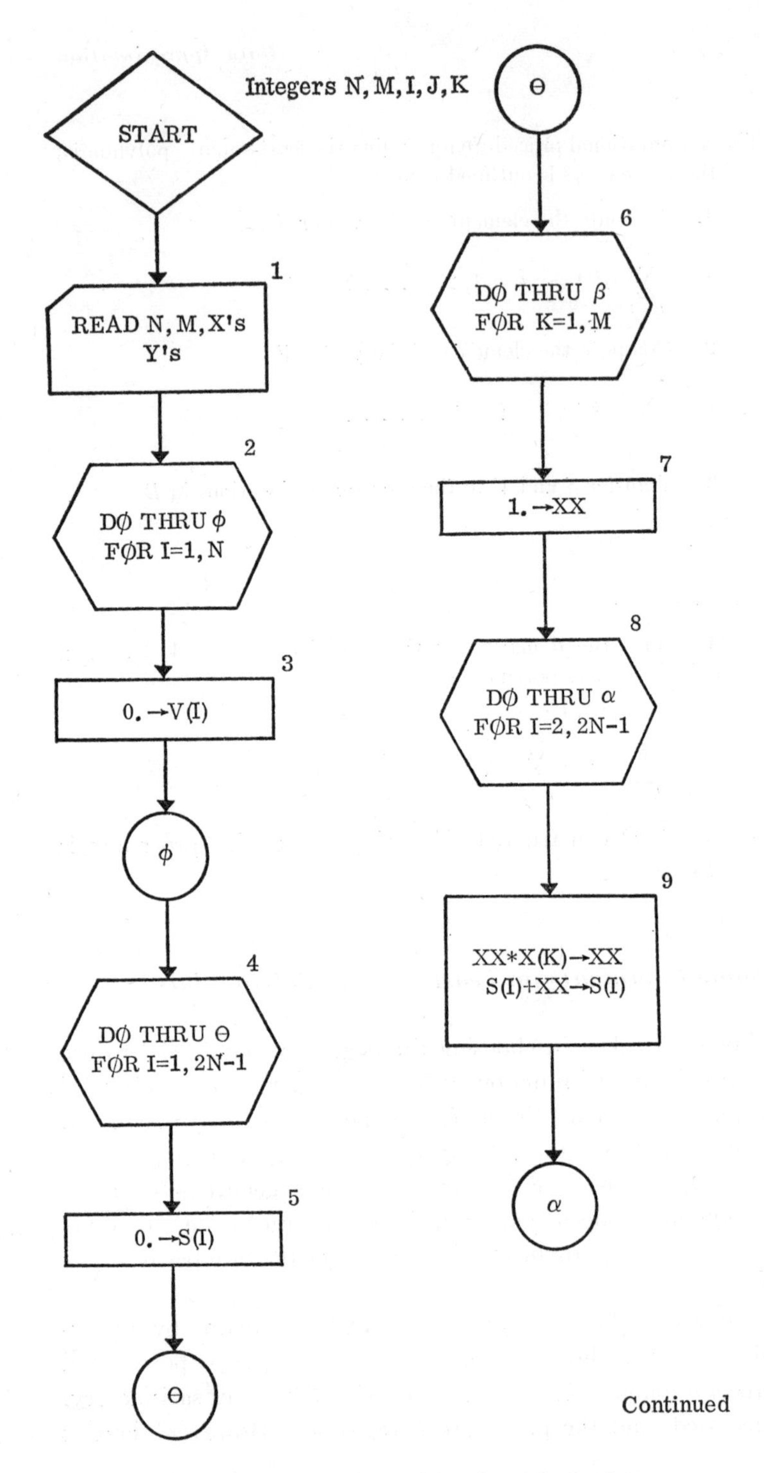

Figure 8.5 Least-squares polynomial to fit a set of points.

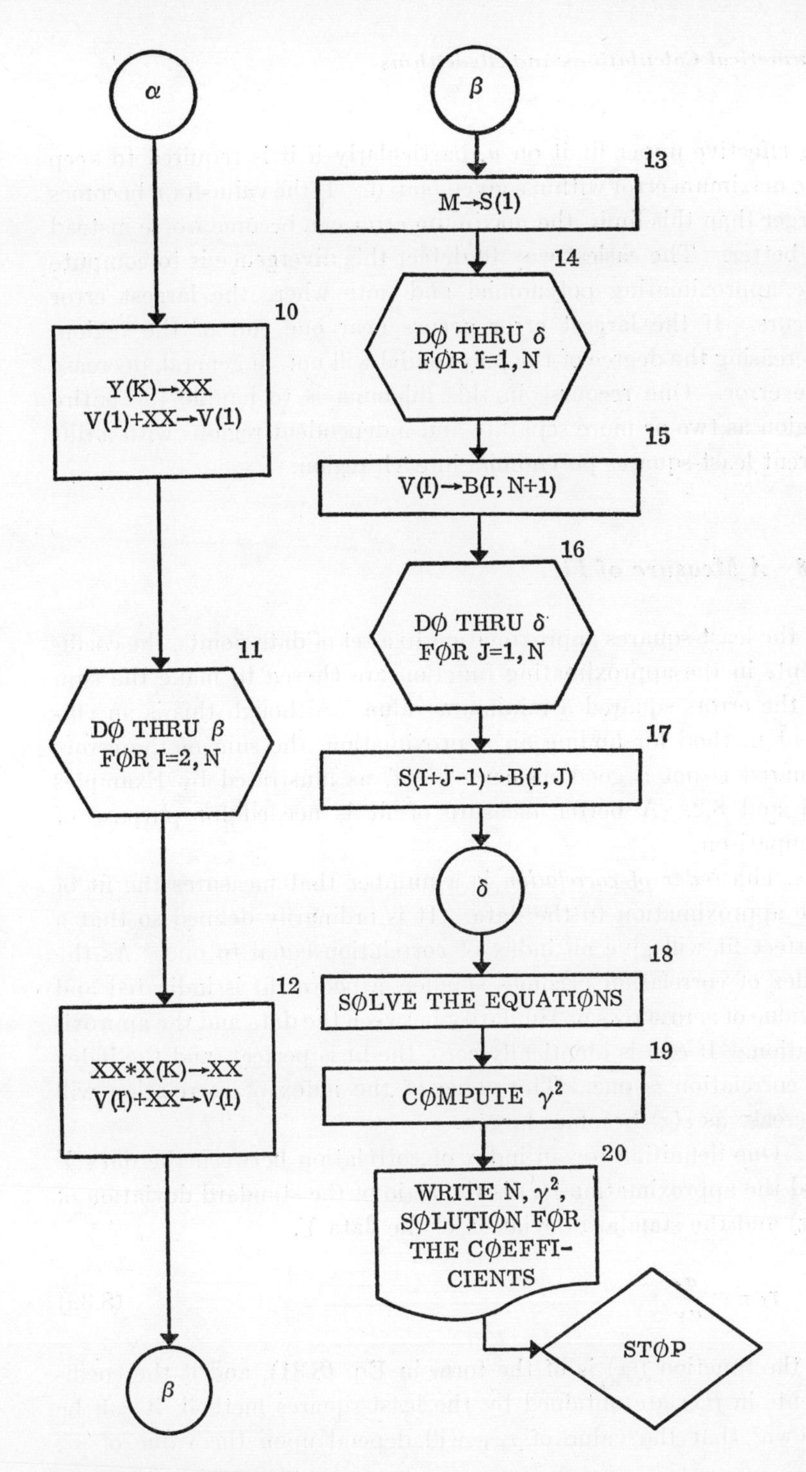

Figure 8.5 *(Continued)*

an effective upper limit on n, particularly if it is required to keep the maximum error within a given bound. If the value for n becomes larger than this limit, the maximum error can become worse instead of better. The easiest way to detect this divergence is to compute the approximating polynomial and note where the largest error occurs. If the largest error occurs near one end of the region, increasing the degree of the polynomial will not, in general, decrease the error. One recourse in this dilemma is to handle the entire region as two or more separate and independent regions with a different least-squares polynomial in each region.

8.8 A Measure of Fit

In the least-squares approximation to a set of data points, the coefficients in the approximating function are chosen to make the sum of the errors squared a minimum value. Although this is an efficient method for finding an approximation, the sum of the errors squared is not a good measure of fit, as illustrated by Examples 8.1 and 8.2. A better measure of fit is needed for purpose of comparison.

The *index of correlation* is a number that measures the fit of the approximation to the data. It is ordinarily defined so that a perfect fit will give an index of correlation equal to one. As the index of correlation becomes smaller, a poorer fit is indicated and a value of zero shows no similarity between the data and the approximation. If $e(x)$ is identically zero, the fit is perfect, and the index of correlation is one. The value of the index of correlation will decrease as $e(x)$ becomes larger.

One definition for an index of correlation between the data Y and the approximation $f(x)$ is the ratio of the standard deviation of $f(x)$ and the standard deviation of the data Y.

$$r_{f,Y} = \frac{\sigma_f}{\sigma_Y} \tag{8.39}$$

If the function $f(x)$ is of the form in Eq. (8.31), and if the coefficients in $f(x)$ are obtained by the least-squares method, it can be shown[3] that the value of $r_{f,Y}$ will depend upon the value of γ^2, according to

$$r_{f,Y} = \sqrt{1 - \frac{\gamma^2}{m\sigma_Y^2}} \tag{8.40}$$

γ^2 is defined in Eq. (8.12), and m is the number of points in the collection Y.

The value of $r_{f,Y}$ is independent of the number of points in the collection. It is equal to one when the error function is identically zero and decreases as the difference between $f(x)$ and Y increases. A value for $r_{f,Y}$ near to one indicates a good fit for the approximating function. The question of how near to one $r_{f,Y}$ should be for a suitable approximation depends upon the problem. For controlled experiments in the physical sciences it is generally expected that the approximation should have a high correlation with the data. In other fields this may not be the case.

Reference is made to Examples 8.1 and 8.2 which give two different approximations to the same function. The coefficients in the two linear functions are nearly the same, and the correlation coefficients are likewise nearly equal, for Example 8.1 $r_{f,Y} = 0.989$ and for Example 8.2 $r_{f,Y} = 0.992$, which shows good agreement between the data and approximation for both.

The methods of this chapter for reducing experimental data are illustrated with a typical laboratory problem.

Example 8.5

The elastic constant μ for an aluminum strip is to be determined experimentally, according to the setup shown in Fig. 8.6.

The constant μ will be evaluated from

$$\mu = \frac{\rho M}{I} \tag{8.41}$$

where ρ is the radius of curvature of the strip on AB, M is the bending moment on AB and is equal to Fa, and I is the moment of inertia of the cross section of the strip about the bending axis. The force F is increased by increments, and the deflection δ is recorded. The data are given in Table 8.2. The values for ρ and M for each value of F are computed from the data according to the equations

$$\rho = \frac{h^2 + 4\delta^2}{8\delta} \qquad M = aF \tag{8.42}$$

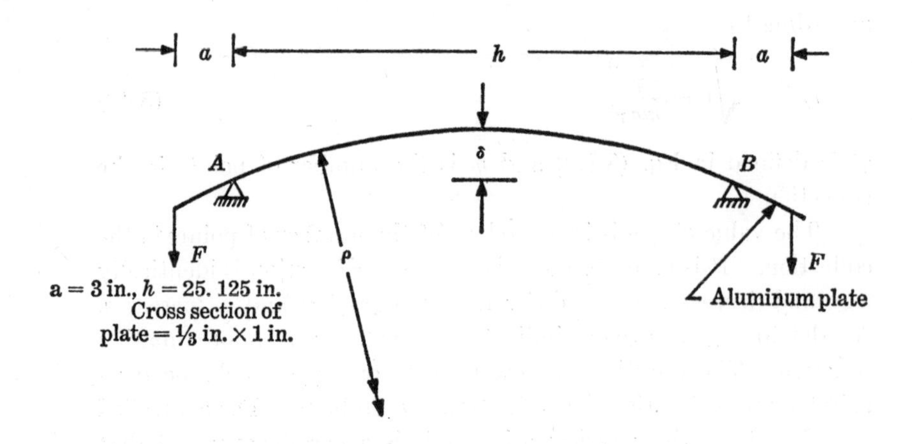

Figure 8.6 Experimental setup for bending a bar.

The product ρM is then calculated, and the values ρ, M, and ρM are entered in Table 8.2.

Table 8.2
Data for Example 8.5

Run no.	F, lb.	δ, in.	ρ, in.	M, lb-in.	ρM lb-in.^{2}
1	0.30	0.0369	2,405.8	0.90	2,165.2
2	0.60	0.0752	1,093.0	1.80	1,967.4
3	0.80	0.1026	805.25	2.40	1,932.6
4	1.10	0.1404	576.90	3.30	1,903.8
5	1.30	0.1635	484.20	3.90	1,888.4
6	1.60	0.2007	392.49	4.80	1,884.0
7	1.80	0.2266	347.73	5.40	1,877.7
8	2.10	0.2657	297.46	6.30	1,874.0

If, as expected, all values for ρM were the same any entry could be used to find the constant μ. This is not the case, however. The value of ρM is monotonically decreasing with F and appears to be approaching an asymptote as F and M get large. The limiting value of ρM as M increases is assumed to be the value that will give the correct value for μ. This suggests an approximation to ρM of the following form:

$$f(M) = c_1 + \frac{c_2}{M} + \frac{c_3}{M^2} + \cdots \tag{8.43}$$

In the least-squares equations the values for ϕ_i, $i = 1, 2, \ldots,$ are given as

$$\phi_1 = 1 \qquad \phi_i = \frac{1}{M^{i-1}} \qquad i = 2, 3, \ldots, n$$

Calculations for the coefficients in the function f, for γ^2, and for the correlation coefficient between f and the data are made for two and three terms in the function f.

For the first approximation, two terms are used.

$$f(M) = c_1 + \frac{c_2}{M}$$

The coefficients c_1 and c_2 are obtained from Eqs. (8.22) and (8.24). The value for γ^2 is found from (8.27), and finally the correlation coefficient is computed from (8.40). The results are given in the first row of Table 8.3 ($n = 2$).

Table 8.3
Solution for Example 8.5

n	c_1	c_2	c_3	γ^2	r
2	1,814.08	306.87		738.63	0.989
3	1,845.75	152.07	121.99	70.59	0.999

A similar set of computations is given for the second approximation, in which the first three terms are used in the function in (8.43) ($n = 3$). The second approximation to the data is excellent. The low value for γ^2 and the very high correlation coefficient verify this. On the basis of this approximation the limiting value for ρM is set equal to c_1. Then

$$\mu = \frac{c_1}{I} = 11.34 \cdot 10^6$$

8.9 Summary

Methods of collocation and least squares are presented for finding a functional relationship between variables in a data record. Comparison of different functions with approximate data is made by means of a correlation coefficient. Several example problems and flow charts are included to illustrate methods of computation.

Problems

1. Each of n samples contains m data. Write a flow chart to find the mean value and the standard deviation of each sample. Let the ith sample be designated $X_i = x_{i,j}$, $j = 1, 2, \ldots, m$.

2. The following sets of data are given:

(a)

p	−0.3	−0.2	−0.1	0.0	0.1	0.2	0.3
q	1.25	1.17	1.08	1.01	0.91	0.83	0.75

(b)

p	0.1	0.2	0.3	0.4	0.5	0.6
q	1.44	1.67	1.98	2.51	3.34	5.00

(c)

p	10	12	14	16	18	20	22
q	0.499	0.996	1.488	1.973	2.450	2.913	3.360

(d)

p	−0.5	−0.3	−0.1	0.1	0.3	0.5	0.7
q	−0.695	−0.359	−0.107	0.097	0.264	0.408	0.533

Find the linear function which best approximates the data.

3. Write a flow chart to solve all of the cases in Prob. 2.

4. What is the index of correlation for the linear function found in Prob. 2a?

5. Write a flow chart to do Prob. 4.

6. The following sets of data are given:

(a)

u	0.1	0.2	0.3	0.4	0.5	0.6	0.7	0.8
v	5.34	5.58	5.67	5.72	5.73	5.75	5.76	5.77

(b)

u	2.01	2.47	2.93	3.39	3.86	4.32	4.79	5.25
v	20.5	19.9	19.7	19.6	19.5	19.5	19.4	19.4

(c)

u	0.00	0.25	0.50	0.75	1.00	1.25	1.50
v	0.200	0.198	0.198	0.197	0.196	0.193	0.184

(d)

u	0	1	2	3	4	5	6	7	8	9
v	3.04	2.88	2.75	2.64	2.55	2.47	2.41	2.36	2.32	2.29

In each case v should be a constant, but because of errors in the experimental setup it is not. The desired value is the limit of v as $u \to \infty$. By the least-squares method, find an approximation to the limit for each set of data.

7. Find the index of correlation for a three-term approximation to the data in Prob. 6a.

8. Write a flow chart to solve Prob. 7.

9. Find the least-squares approximation to $y = \sqrt{x}$, where $f(x) = c_1 + c_2/x$. Use $x_j = 0.1j$ for $j = 1, 2, \ldots, 10$ as data points. Graph $y = \sqrt{x}$ and the approximation $f(x)$.

10. Do Prob. 9 for the continuous case on $0.1 \leq x \leq 1.0$.

11. Write a flow chart for finding the least-squares linear approximation for a sample of data represented by $x_i, y_i, i = 1, 2, \ldots, m$.

12. Construct a flow chart to find the linear least-squares approximation for each of n sets of data represented by (X_i, Y_i). $X_i = x_{i,j}$, $Y_i = y_{i,j}$, $i = 1, 2, \ldots, n$, $j = 1, 2, \ldots, m$.

13. Write a flow chart to do Prob. 12 with the addition that the linear correlation coefficient is computed.

14. Construct a flow chart to give the index of correlation between the points on the curve $y = e^x$ for $x = 0.1j$, $j = 0, 1, \ldots, 10$, and the line $f(x) = x$.

15. A sample of data represented by $X = x_j$, $Y = y_j$, $j = 1, 2, \ldots, m$, is approximated by the curve $y = a \sin x + b \sin 2x$. Construct a flow chart to find the value of a and b for the best least-squares approximation. Include in the flow chart a computation for the index of correlation.

16. A function f is dependent upon two variables x and y. For a sample of data x_i, y_i, $f(x_i,y_i)$ for $i = 1, 2, \ldots, n$, construct a flow chart to give the best linear approximation to the data. *Note:* Find by least-squares approximation the value of a, b, and c in $f(x,y) = a + bx + cy$ such that the sum of the squares of the error between the function and the data is minimized.)

17. Write a flow chart for evaluating the coefficients in the least-squares polynomial of third degree to approximate a set of data $(x_i, y_i, i = 1, 2, 3, \ldots, m)$. Include all steps in the computation routine.

18. Write a routine for finding the index of correlation for Prob. 16.

References

1. Hamming, R. W.: "Numerical Methods for Scientists and Engineers," McGraw-Hill Book Company, New York, 1962.

2. Hildebrand, F. B.: "Introduction to Numerical Analysis," McGraw-Hill Book Company, New York, 1956.
3. Hurt, James: "Some Numerical Techniques for Use on a Digital Computer," unpublished M.S. thesis, Department of Mechanics and Hydraulics, University of Iowa, 1963.
4. Jackson, D.: "Fourier Series and Orthogonal Polynomials," Mathematical Association of America, 1961.
5. Ralston, Anthony: "A First Course in Numerical Analysis," McGraw-Hill Book Company, New York, 1965.

Index

Index